The Chemistry of Photography

From Classical to Digital Technologies

Dedicated to
my father
Joseph Rogers

The Chemistry of Photography

From Classical to Digital Technologies

David Rogers
Danercon Ltd., Harrow, Middlesex, UK

RSC Publishing

ISBN-10: 0-85404-273-3
ISBN-13: 978-0-85404-273-9

A catalogue record for this book is available from the British Library

Published by The Royal Society of Chemistry,
Thomas Graham House, Science Park, Milton Road,
Cambridge CB4 0WF, UK

Registered Charity Number 207890

For further information see our web site at www.rsc.org

Typeset by Macmillan India Ltd, Bangalore, India
Printed by Henry Lings Ltd, Dorchester, Dorset, UK

Preface

Critics of the use of silver halide in photography would have us all believe that the whole world has now moved over to digital systems, and that the old days have gone forever. While it is certainly the case that the market for the use of silver halide is in decline and may continue to decline for some years yet I, for one, would suggest that the wet chemistry approach to photography will not cease completely – at least for many years to come.

Digital photography has indeed passed the stage where only the new adopters have embraced the technology, for it is now a practical reality in most homes, and indeed on most new mobile telephones. Yet, there are still enthusiasts who will continue to use the chemical means of creating images at least for the foreseeable future. Additionally, silver halide colour photographic papers can be used to print digital images. These images have exceptional image stability and may provide a vital part of the overall photographic experience for the next few years.

If we assume for a moment that we are in the transition between the technologies, and that silver halide/wet chemistry photography will indeed terminate at some point in the near future, perhaps the time is right to review the technology, which was developed and largely taken for granted by most of the photographic consumer market. For comparison purposes, the chemistry of one of the photographic papers used to print digital images and the chemistry of inkjet paper is also included.

It is a remarkable technology. Amateur films can contain up to 100 distinct chemicals, coated in very low or indeed no light levels. The silver halide system works at the molecular level and therefore works in the millions of pixels per inch. Silver halide prints are carried and viewed by millions of people and adorn millions of walls as display items, all without the need for any power whatsoever except for viewing, which can take place using the natural energy of the sun. At its peak, the industry turnover for silver halide products was calculated in the late tens of billions of dollars and affected all of the inhabited continents around the world.

This overview is but a fraction of the total knowledge that has been generated by the large photographic manufacturers over the last 100 years. It is intended for students to reflect and discover the complexity of the chemistry that many have taken for granted. Perhaps the photographic system, more than any other in the technological world, has combined the use of organic, inorganic and physical chemistry with elements of engineering and physics. Many texts exist that have covered many aspects of the overall system with the possible exception of the chemistry. This volume seeks to address this oversight.

A volume such as this requires input from many sources. The author was privileged to work for the world's largest photographic company for over 20 years splitting his time between the research and development communities and the manufacturing division. During that time, he met many people who gave of their time in explaining the inner workings of this complex technology. Of particular note were his conversations for organic chemistry with Joe Bailey, Dave Clarke, Trevor Wear, Alan Pitt, Judith Bogie; film design with Gary Einhaus, Drake Michno, Paul Magee, John Higgins, Mike Simons, Alan Eeles; processing chemicals John Fyson and Peter Twist; emulsion techniques Roger Piggin, Adrian Codling and Gary Hiller; discussions concerning graphic arts film technology Bill Fardell and Tim Peachey. This august body of scientists was not the only source of knowledge, but these people were fundamental in the author's training and development with the exception of one individual. That individual, John Sawyer, once gave the author a 'bearded eagle' award for attending all of his lectures concerning film design. John Sawyer was perhaps unique in his ability and willingness to pass on his knowledge and experience. The author is forever grateful for the time John willingly provided, and for the bearded eagle award.

The more practical aspects of assembling this text also involved a number of people, including Ziaad Khan of the British Library, the staff at the library and information centre of the Royal Society of Chemistry, the library staff of the Royal Society and the University of Westminster.

My thanks also to Kate Price at Country Ways and to Peter Whitfield who provided some timely help with some software.

Finally my family, Carolyn, Adam and James, provided the space and time for me to lock myself away to produce this text. Thank you one and all.

Contents

SECTION 1: CONVENTIONAL FILMS AND PAPERS

Section 1: Conventional Films and Papers

CHAPTER 1

The Overall System (Capture and Output)

The use of colour filters in the production of colour images pre-dates colour photography as we know it by almost 80 years. James Clerk Maxwell produced coloured images in 1861 by projecting the same image through three projectors, each of which had one of the three colour filters of red, green and blue. Maxwell produced silver positive images, which were created from emulsions that were not spectrally sensitised. These images were surprisingly good. They relied on the additive colour system, whereby the three primary colours (red, green and blue) produce the full gamut of colours on the composite final image, Figure 1.

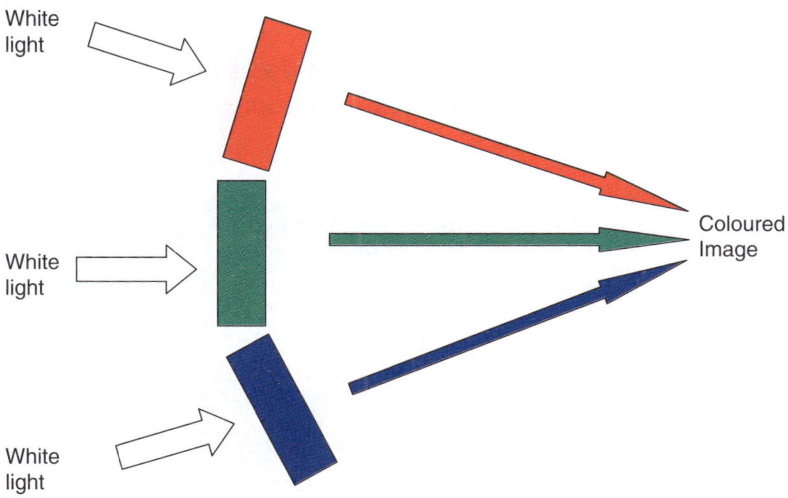

Figure 1 *Maxwell's early experiment*

1.1 The Additive Colour Process

Each filter was designed to transmit light of only one primary colour. This system might find limited application in transmission systems where three images can be carefully brought together in register. Unfortunately, this system is cumbersome. It is possible to produce a coloured image but it is very difficult to align the three colour images, certainly with any speed. The combination of the three filters into one pack, which would allow for image alignment, is not a practical option either as there would be no light transmitted through the pack under some circumstances, especially if the filters were to have much colour density to them, Figure 2.

A more practical alternative, at least in terms of image alignment, is to expose and view all the colour images at the same time. Under these circumstances, it is not possible to use the additive system as described above, for the obvious fact that there would be no light exiting the images.

The subtractive colour process, on the other hand, subtracts either red, green or blue light from the visible spectrum *i.e.* each dye subtracts one third of the visible spectrum and not two thirds, as is the case for the primary colours. Under these circumstances the original image produces a positive, through the intermediate step of a negative image, Figure 3.

1.2 The Subtractive Colour Process

The subtractive colour system uses the dyes cyan, magenta and yellow. The combination of cyan and magenta produces blue, yellow and cyan produces green and magenta and yellow produces red. A white positive

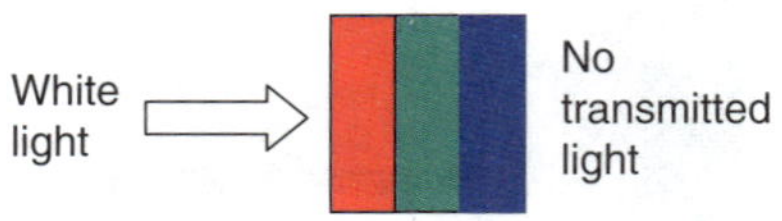

Figure 2 *Little or no transmitted light*

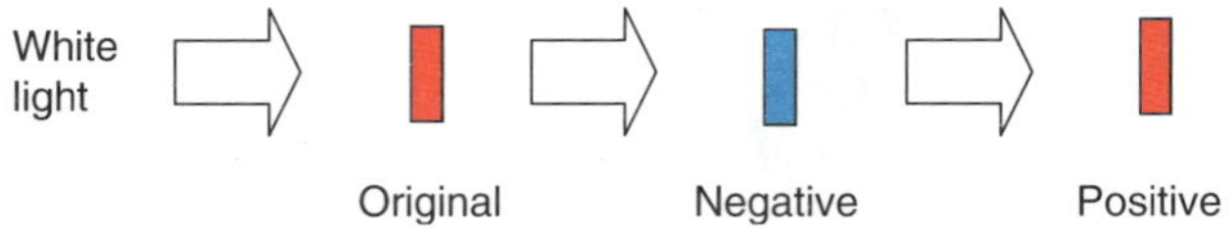

Figure 3 *The subtractive colour process*

image is achieved by producing a totally black negative, *i.e.* exposure of all three of the cyan, magenta and yellow records (in the negative), whereas a black positive image results in the absence of colour in the negative.

In practice, the situation is slightly more complex as there are some physical constraints. Silver halide crystals are inherently sensitive to blue light and so a yellow filter layer needs to be introduced between the yellow records and the other two records, making layer order important. Silver halide crystals need to be spectrally sensitised so that they capture as much of the available light as possible. The cyan, magenta and yellow dyes are not perfect dyes and therefore there is a need for colour correction. In addition, the initial material needs to contain, at least as far as possible, colourless components during the exposure of the negative. Additionally light may simply be reflected either between the silver halide grains or the various surfaces. Also the photographer may require or need exposure latitude, so that he/she can expose a scene under a variety of light levels. The exposed film may rest in a glove compartment of a car where the temperatures may rise somewhat. It may also be the case that the film is used to take pictures during Christmas holidays, and the film not used until the next year. Alternatively, the amateur photographer may try to use one film for both winter and summer shots, where the lighting conditions are completely different. The negative film therefore needs to be stable to humidity/light, *etc.* and have the capability of retaining the latent image for a long period of time.

These considerations aside, at least for the moment, the overall system of positive and negative, might be described in Figure 4.

The positive image displayed above may not, at first glance, look like a copy of the original scene. Perhaps it is easier to consider an actual scene. The picture below (Figure 5) is a yellow image of an actual scene. On close inspection, one can see that it is not entirely yellow, there is contamination from other colours. This is because the yellow dyes are not pure and have unwanted absorptions, which will be covered in detail in Chapter 8.

The corresponding cyan image is shown in Figure 6.

These images, when combined together, produce a green reproduction of the original scene, Figure 7.

Similarly, magenta (Figure 8) and cyan (Figure 9) images together form a blue image, Figure 10.

Finally the combination of yellow (Figure 11) and magenta (Figure 12) produces a red image, Figure 13.

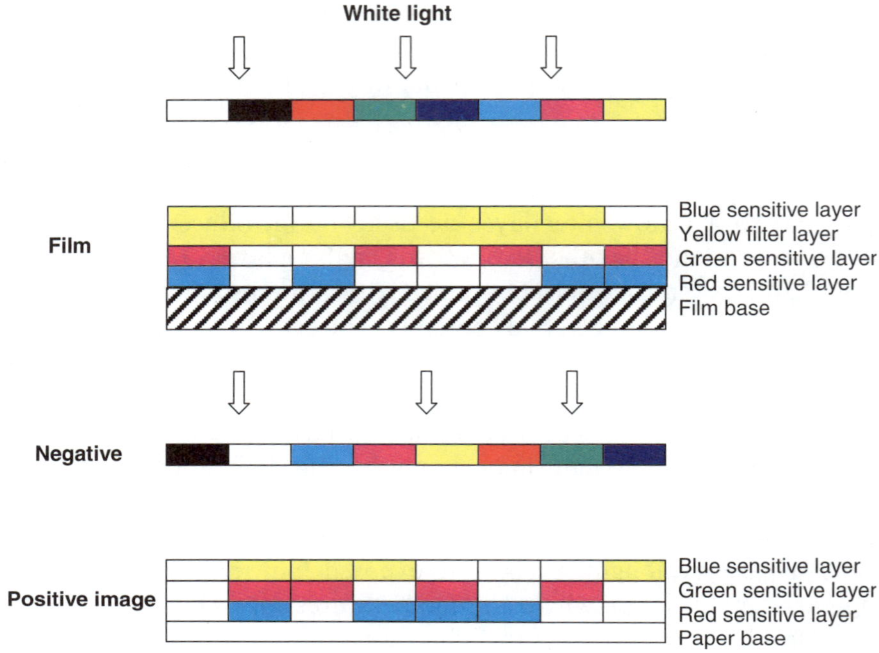

Figure 4 *The subtractive colour process for all film layers*

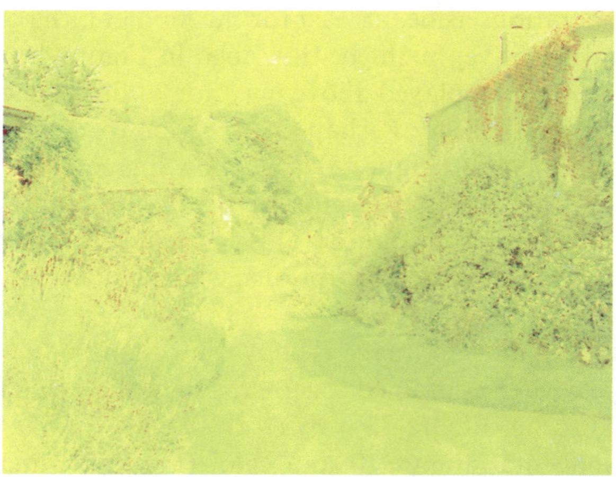

Figure 5 *Yellow record of an original scene*

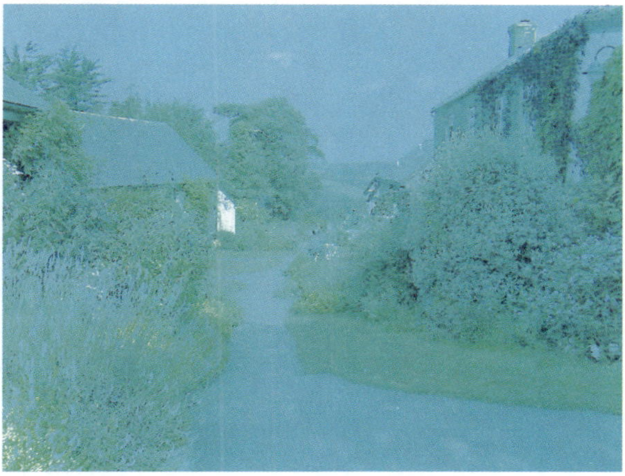

Figure 6 *Cyan record of an original scene*

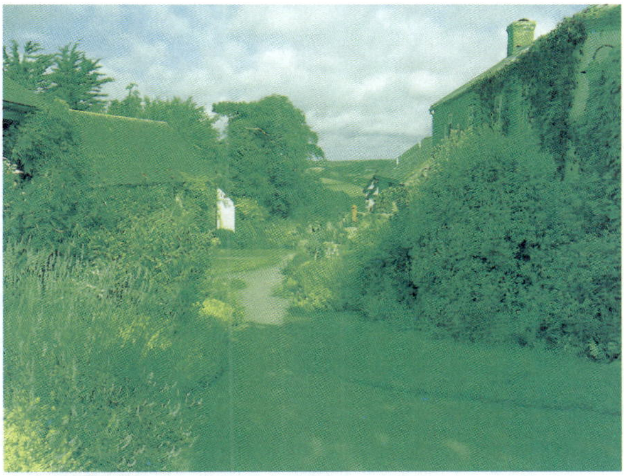

Figure 7 *Green image produced by combining the yellow and cyan records*

The combination of cyan, magenta and yellow returns the original scene, Figure 14.

The two-stage process of negative and positive provides the photographic manufacturer with the means to convert a positive image to an intermediate negative, and then back to the positive image again in the photographic paper. The use of the subtractive colour process with transparency film is somewhat more complex, as there is no physical intermediate. Transparency films are designed to be projected onto a screen. So how can a transparency film use the subtractive process and, without the use of a physical intermediate, afford a positive image capable of being viewed by projection?

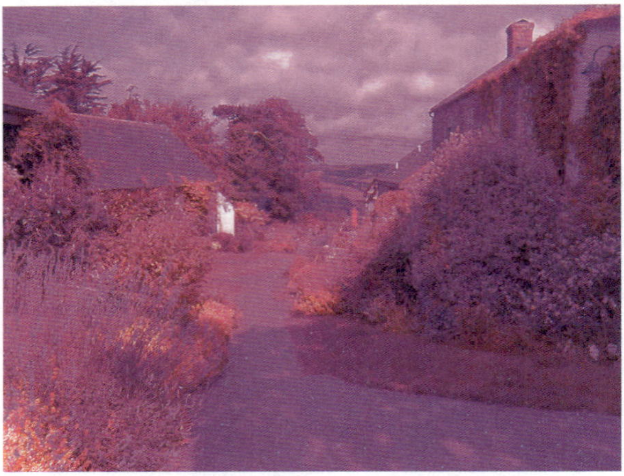

Figure 8 *Magenta record of an original scene*

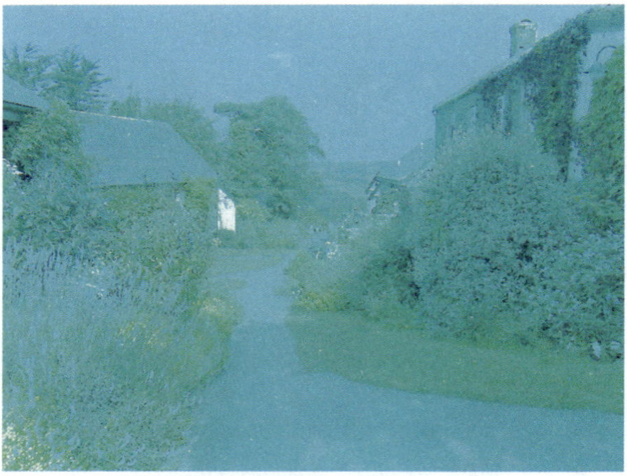

Figure 9 *Cyan record of an original scene*

The method used is to generate an intermediate negative within the photographic layers, during the processing stage, and to subsequently reverse the image during the processing stages.

The schematic diagram below shows the process, Figure 15.

This reversal process is unusual from many perspectives, see Chapter 11. While it is possible to evaluate pictures against criteria such as:

- How sharp is the film?
- How much photographic latitude does it have (the ability for under and over-exposure)?
- How grainy is the scene?

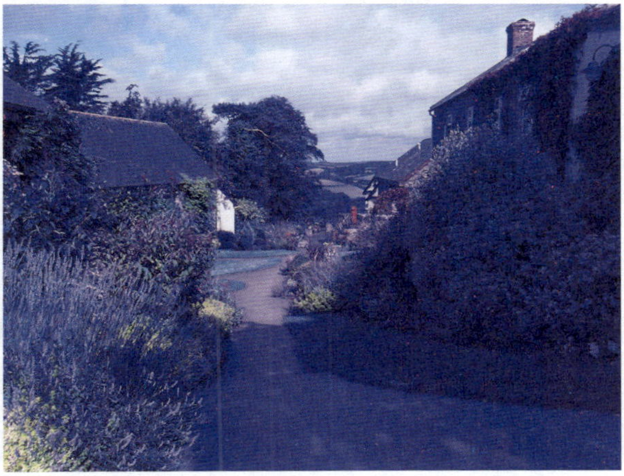

Figure 10 *Blue image produced by combining the magenta and cyan records*

Figure 11 *Yellow record of an original scene*

The more rigorous approach is to expose the film or paper to various test objects under standard lighting conditions, so that objective and quantitative photographic parameters can be measured, and several competitor's products or experimental films and papers compared objectively.

On first inspection, one might assume that there is a simple relationship between the exposure given to a photographic film or paper and the subsequent density produced. The relationship between exposure and

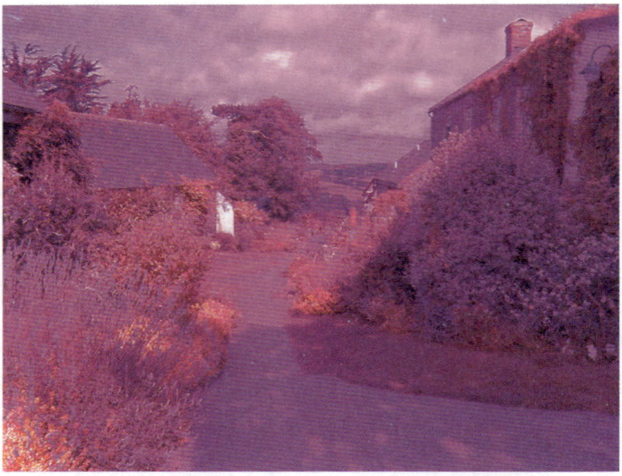

Figure 12 *Magenta record of an original scene*

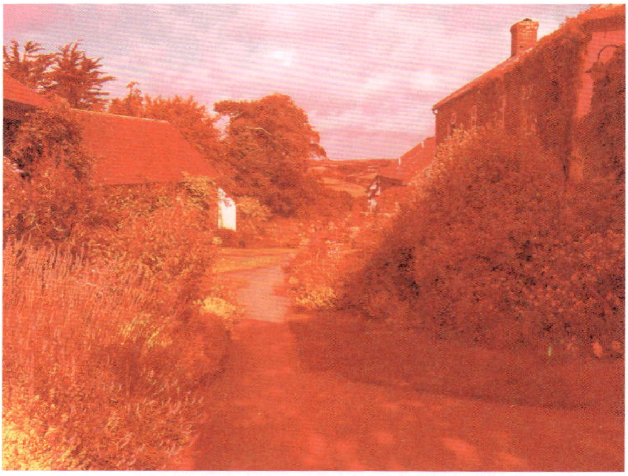

Figure 13 *Red image produced by combining the yellow and magenta records*

density is, however, more complex. In practice, density is plotted against the log of the exposure. A simplified schematic of a test object, which is commonly used to generate a density *vs.* log exposure curve is given below, Figure 16.

A white light exposure of a colour film exposes all three colour records, producing this graded image, more commonly known as a 'grey scale'. In this case each distinct density area will produce a density reading in each colour record, as it needs three density records to produce a black negative image. The step wedge densities are known

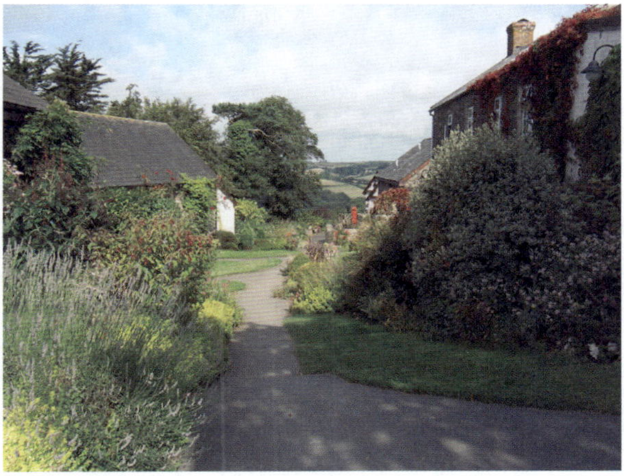

Figure 14 *The original scene*

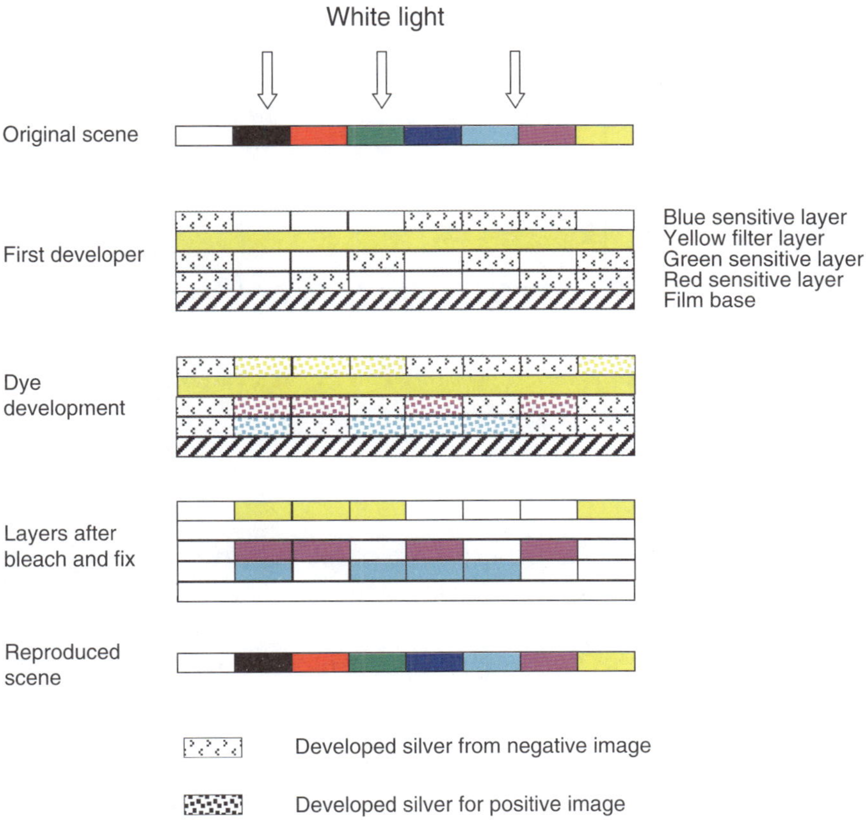

Figure 15 *The colour reversal process*

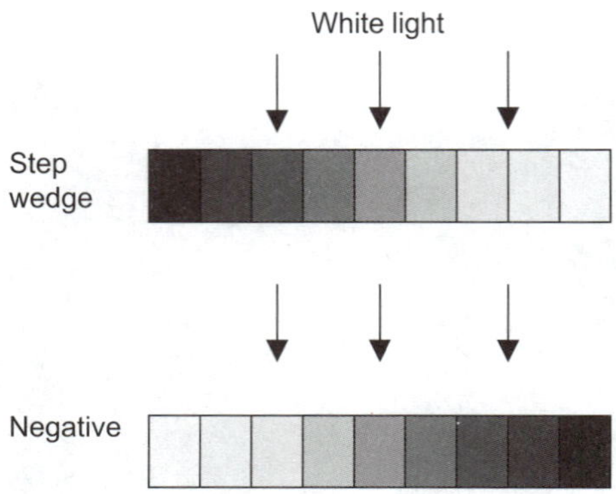

Figure 16 *Negative image produced by a step wedge*

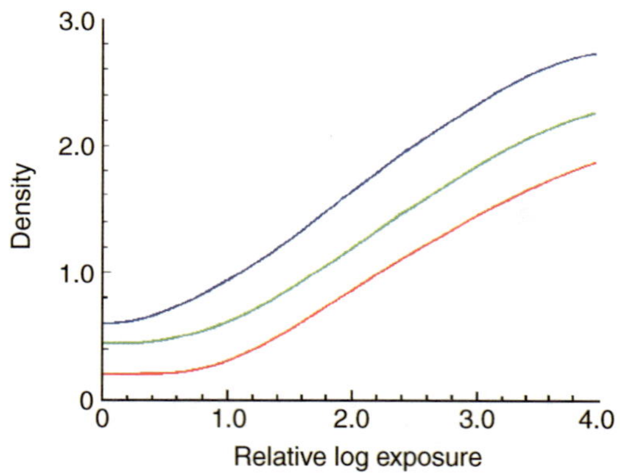

Figure 17 *Typical colour negative film density v log exposure curves*

to a high accuracy, and so the exposure of the negative for each step on the wedge can be plotted. The density steps on the colour film or paper will, when plotted against the exposure, produce three curves. One example of the type of curves produced from a colour negative film is given below as Figure 17.

The straight-line portion of the curve is known as the contrast or gamma. It is different for different applications. In this particular case the contrast may be in the region of 0.7, in graphic arts films the contrast might be as high as 5.0–6.0, because that application requires the formation of dots and not a continuous image. The lower part of the curve is known as the 'toe', and the upper part of the curve is known as the 'shoulder' or 'D_{max} region'. This is the maximum density that is possible to create with the combination of exposure/dye and silver laydown levels used for that formulation. The relative separation of the dye curves with exposure provides the filmbuilder with the challenge of creating consistent colours in a scene, if photographed under different lighting conditions. This issue is extremely important and will be covered in the discussion concerning the chemistry of colour, see Chapter 8. The density/log exposure or sensitometric curves for colour films and papers appear in many of the standard texts and are discussed in various subsequent chapters.

Graphic arts films require high contrast as the method of generating the image with graphic arts films is to generate a halftone image, which is made up of dots, the size of which determines the level of light and dark in the picture. For example, Figure 18 shows a uniform halftone dot.

For demonstration purposes, Figure 19 shows our standard scene as a halftone image.

A close up of the image shows the halftone dots, see Figure 20.

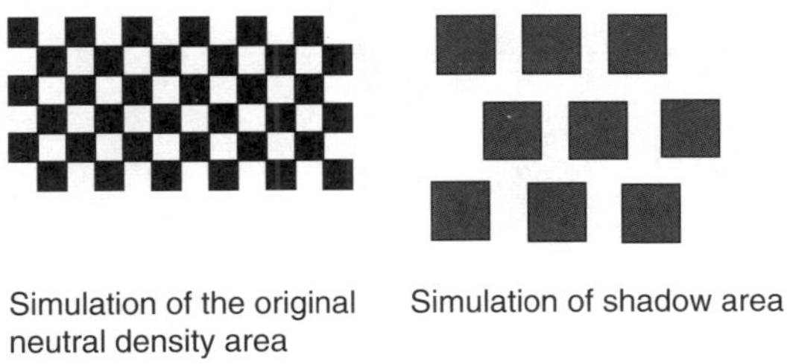

Figure 18 *Uniform halftone dots*

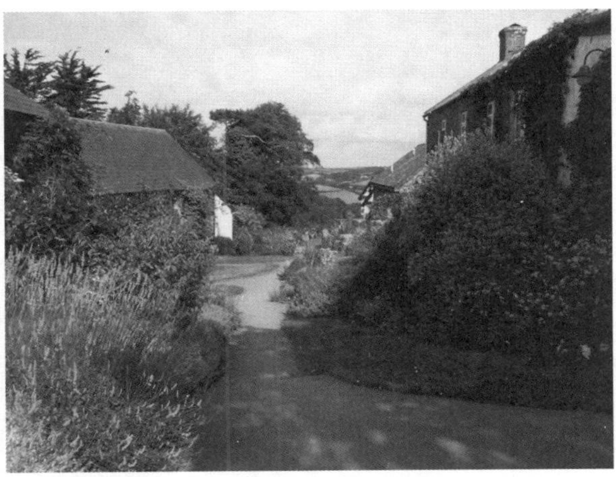

Figure 19 *Standard scene produced using halftone dots*

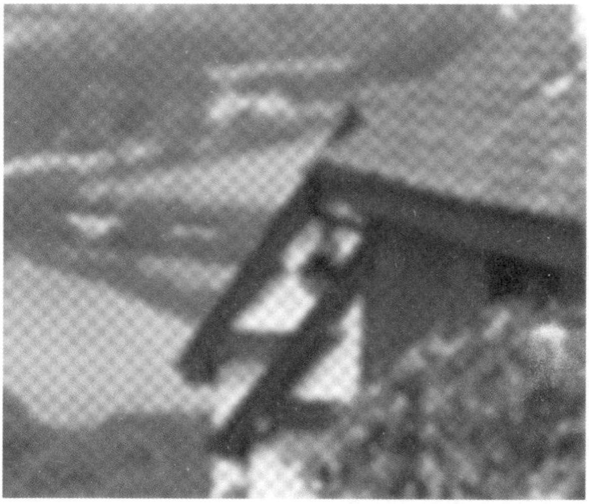

Figure 20 *Close up of part of the standard scene shown as a halftone*

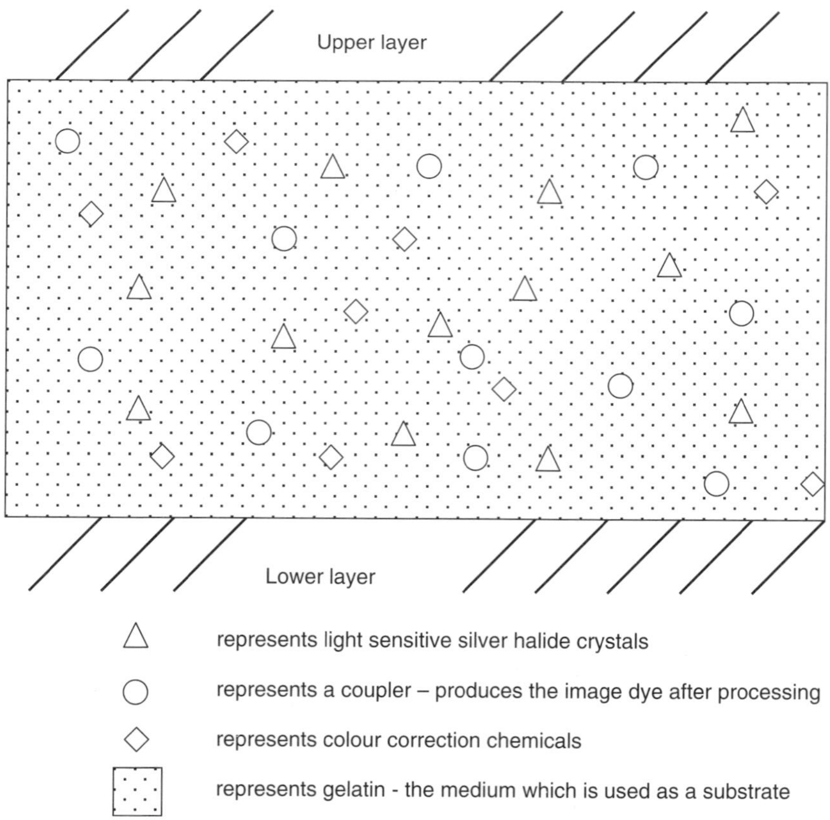

Figure 21 *Some typical components coated in a colour negative film layer (not to scale)*

Photographic components can be application-specific, for example black and white films do not use colour couplers. Chapter 6 discusses these compounds. Similarly photographic paper products do not use clear plastic base as films do. The most complex photographic product is colour film. Prior to a discussion of the chemistry of this or indeed any other product, it is worth considering the type of chemicals that are coated in an average colour negative film, Figure 21.

1.3 Cross-Section of a Typical Colour Film Layer

The size and number of shapes have been used for demonstration purposes only. They do not represent the relative amounts of the various chemicals coated during the manufacturing process.

The light sensitive silver halide crystals capture the light when film is exposed. During processing the latent image is magnified, oxidising the colour developer. This then reacts with the couplers to form dye in the image areas, or reacts with image correction chemicals. Gelatin is the coating medium for all of these chemicals.

The types of chemicals listed above are not the only ones coated. There is a range of coating aids, usually coated in one layer (but capable of diffusion to all of the layers) and hardener to cross-link the gelatin, *etc.* During the next few chapters, each of the various components will be described in more detail.

CHAPTER 2

Gelatin

Gelatin has been the medium of choice in which to disperse and coat photographic materials for well over 100 years. Over that time there have been many research projects that have looked at potential alternatives. For one reason or another, all of the synthetic alternatives have properties or costs, which prohibited their use as a photographic medium. While the properties of gelatin will be covered in more detail later in this chapter, it is worth recording here that gelatin is a naturally occurring substance and is a bi-product of other industries.

Gelatin is extracted from collagen, which is the most abundant protein of the higher mammals, being present in connective tissues such as bone, cartilage, ligaments, skin and tendons. Gelatin can be extracted from collagen using either basic or acidic conditions, the residue being lipids, mucopolysaccharides, non-collagen protein and polynucleic acids.

The Gelatin Manufacturer's Institute of America (established 1956) was formed to *carry on research in the manufacture and usage of gelatin and to carry on promotional work in its uses*. Representing the interests of gelatin manufacturers across North America they published the protein quality of gelatin on their website (http://www.gelatin-mia.com/html/gelatine_health.html), some data from which is reproduced below (Figure 1). It is unclear if these results are from the analysis of acid or alkali washed gelatin. This data, however, records the types and amounts of amino acids present in gelatin – at least in the higher animals.

The multi-stage process of converting bones into gelatin is detailed on the websites of various suppliers of photographic gelatin, see for example (http://www.eastmangel.com/ (suppliers to Eastman Kodak Co. since 1930) and http://www.rousselot.com/index.html). Gelatin produced from Rousselot is used in the United States and Europe. The incoming raw materials can be from a number of sources, the chemical content of which varies.

Amino Acid	g amino acids per 100g pure protein
Glycine	27.2
Proline	15.5
Hydroxyproline	13.3
Glutamic Acid	11.6
Alanine	11.3
Arginine	9.0
Aspartic Acid	6.7
Lysine	4.4
Serine	3.7
Leucine	3.5
Valine	2.8
Phenylalanine	2.5
Threonine	2.4
Isoleucine	1.6
Hydroxylysine	0.8
Histidine	0.7
Methionine	0.6
Tyrosine	0.2
Tryptophan	0.0

Figure 1 *Typical amino acids found in gelatin*

2.1 The Gelatin Manufacturing Process

Figure 2 outlines the key process variables and steps in the process deemed important in photographic quality gelatin. Eastman Gel reports that it takes about 6 kg of cattle bone to produce each kilogramme of gelatin, which serves to show the level of waste. It is also not a quick process. Even though there are checks throughout the process, some trace element concentrations vary from batch to batch. A few of these trace elements have photographic effects, as they react with or affect the silver halide crystals. These trace elements cannot be removed but their concentration levels can be tolerated if the levels of the impurities are kept constant. In some cases this can only be achieved by blending different gelatin batches.

Other non-photographic industries use gelatin, including baked goods, icing and gelatin desserts. Gelatin is also used as a base for cosmetic and pharmaceutical products, for example the coating on a gel cap style pill. The gelatin produced for these food applications is actually produced to a level of impurities that is higher than for photographic quality gelatin.

Gelatin is a mixture of natural polymers, and the medium into which photographic components are mixed, prior to coating. Once coated into products the layers are hardened by cross-linking the gelatin polymers. The photographic industry uses the following properties of gelatin:

The Gelatin Manufacturing Process

Activity	Key ProcessVariables
Incoming bone	bone moisture
Acidulation	acid concentration
	diluted HCl temperature
Liming	liming time v. aim
	lime carryover
	liming temperature
Washing	wash water iron content
	wash water chlorine content
	lime wash time
	final Stock pH
Extraction	de-ionised water/iron content
Filtration	heat treatment
	residence time
Evaporation	gel concentration
Filtration	
Cooling	finish gelatin pH
	de-ionised gelatin pH
Drying	final moisture
	PA/DI PA moisture
	de-ionised gelatin moisture
Blending	
Shipping	

Figure 2 *Key process steps in gelatin manufacture*

- vertical swell
- optical rotation of light
- hardening/cross-linking capabilities
- impurities
- available viscosity range
- a range or molecular weights
- 'crackability'/the ability to bend.

These have been known for many years since the broad application, *i.e.* a medium in which photographically active materials are dispersed, has not changed from the early days of photography. Indeed, Sheppard and his co-workers[1–3] at Eastman Kodak Co. published some of the seminal work over 70 years ago.

There are several analytical tests to which photographic products might be subjected, which are designed to test/understand coated and dried gelatin layers. They include

- bloom
- gel strength
- melting point
- scratch resistance

- swelling
- wet abrasion, sometimes known as mushiness.

These tests are designed to simulate the uses (and abuses) of commercial products. Samples are also incubated in ovens of varying temperature and relative humidity (perhaps 70°C with a relative humidity of 50%), again designed to test the hardened/cross-linked gelatin layers.

Melting point is the term that is used to describe the temperature at which a gelatin layer will separate from the base upon which it is coated. Gelatin is known to change its properties above 40°C, as it becomes a mix of extremely polydispersed molecules. Below 40°C, however, the coiled gelatin molecules undergo a transition into a helical conformation and resemble stiff rods. The melting point test may reach or exceed this temperature, but as the melting point test is a destructive test, the gelatin structure is not an issue.

These and other properties of gelatin are often reported at international conferences – a series of which ran in the 1970s. The Imaging Science Group of the Royal Photographic Society sponsored a gelatin conference in 2005 – see http://www.rps-isg.org/gelatin2005.php.

Gelatin coatings need care when they are dried as it is possible to affect the coating, particularly if low temperature chilling is followed by high temperature drying. The defect created by such thermal treatment is known as 'reticulation', see the example in Figure 3. The picture below is part of the standard picture, magnified to demonstrate reticulation. It

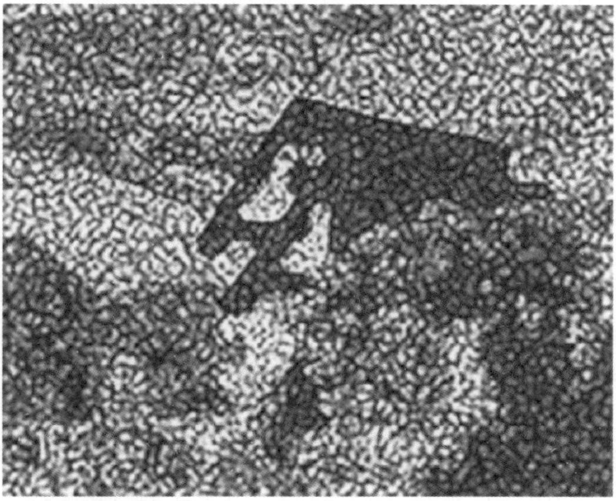

Figure 3 *Simulated reticulation*

is produced in black and white so that the reader is not distracted by any dye cloud issues.

Repeating the drying cycle at more appropriate temperatures with a fresh sample of photographic material should resolve the issue, as would an increase in the hardener level. Hardener levels are strictly controlled during the manufacturing process, and so should not be the cause of reticulation as seen by a customer. Nevertheless, abnormally low hardener levels will affect the film properties. It would also affect the swelling characteristics of the photographic product.

Hardeners may be derived from a number of different chemical families, for example

- Aldehydes, *e.g.*, formaldehyde – higher alkyl homologs of formaldehyde have no effect on gelatin
- Aldehyde acids, *e.g.*, 2,3-dichloro-4-oxo-2-butenoic acid
- Bisaziridines
- Bisepoxides
- Carbodiimides
- Compounds with activated double bonds, *e.g.*, divinylsulfones. N, N^{I},N^{II}-trisacryloylper-hydro-s-triazine
- Dichlorotriazine derivatives, such as 4,6-dichloro-2-hydroxy-5-triazine
- Diketones, *e.g.*, 2,3-butanedione, 1,2-cyclopentanedione
- Dihalides, *e.g.*, 1,3-dichloropropanol
- Diisocyanate bisulfite adducts
- Epoxides
- Isocyanates
- Polybasic acids – specifically anhydrides and acid chlorides
- Sulfonate esters
- Sulfonyl halides *e.g.*, bis(sulfonyl chlorides).

In general, these hardeners reduce swelling in water and increase the melting point of the layer. The effectiveness or degree of hardness is usually measured by a variety of methods

- swell measurements
- determination of the melting point
- viscosity changes
- abrasion resistance.

Many workers have reported on the structure of gelatin, cross-linked particularly with formaldehyde as the hardener,[4] and of hardening in general.[5] In their research paper titled *Rheological Properties of Swollen Gelatin Coatings*,[6] D.J. Taylor and A.M. Kragh report the results on their studies. They divide their results into

- effect of cross-linking
- effect of temperature
- effect of drying conditions
- slow penetration resistance.

The most commonly used hardener for colour negative films is bis-vinylsulfonylmethyl ether, see for example.[7,8] The reported synthesis, 1971,[9] is described in the patent thus

> *... the reaction used for preparing the compound of this invention may be conducted in the presence of organic solvents at moderate temperatures and may be carried out at super-atmospheric or sub-atmospheric pressure. The compound may be prepared by reacting sodium 2-hydroxyethylmercaptide with di-(chloromethyl)ether, oxidizing the thioether atoms to sulphone groups with hydrogen per-oxide, replacing the hydroxyl groups with chlorine atoms in known manner, for instance with thionyl chloride, and then removing hydrogen chloride, for instance with triethylamine, to leave the desired product ...*

One of the properties of cross-linked gelatin significant in the fabrication of photographic papers is the propensity of gelatin to cause internal reflections at the air-gel interface. Gelatin has a refractive index of 1.5, which means that light reflected from the paper base back through the gelatin layer will only emerge if the angle of the light is less than 40° from the perpendicular, Figure 4.

Only 38.6% of the light reflected from the gelatin/paper base surface emerges from the gelatin layer at the first attempt. The remainder of the light will be further reflected and may or may not, finally emerge from the layer. The consequences for the reflection densities of colour paper image dyes is discussed later – see Chapter 6 Figures 1 and 2.

A less common method of producing images using a gelatin substrate relied on producing a gelatin relief image, where the depth of gelatin relates to the intensity of the exposure. Having produced the image, the hardener solution is used to harden the gelatin in the exposure area. Unexposed emulsion is then washed off and the relief image dyed to produce a final image. This type of material was known as the Kodak Dye Transfer (Figure 5).

For the dye transfer process to work effectively, pyrogallol was added to the developer solution and the amount of sulfite (used as an oxidised

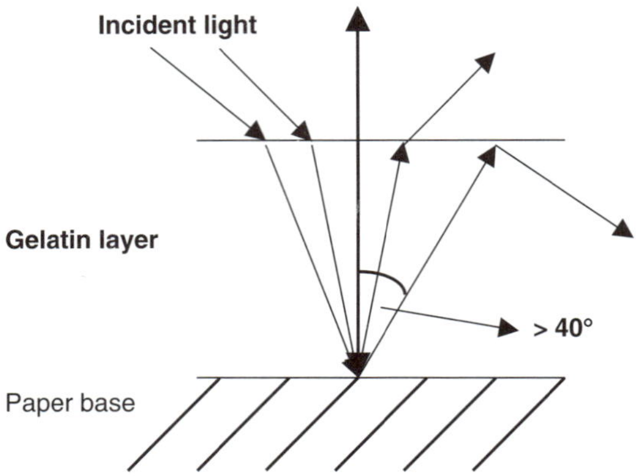

Figure 4 *Internal reflections within a gelatin layer*

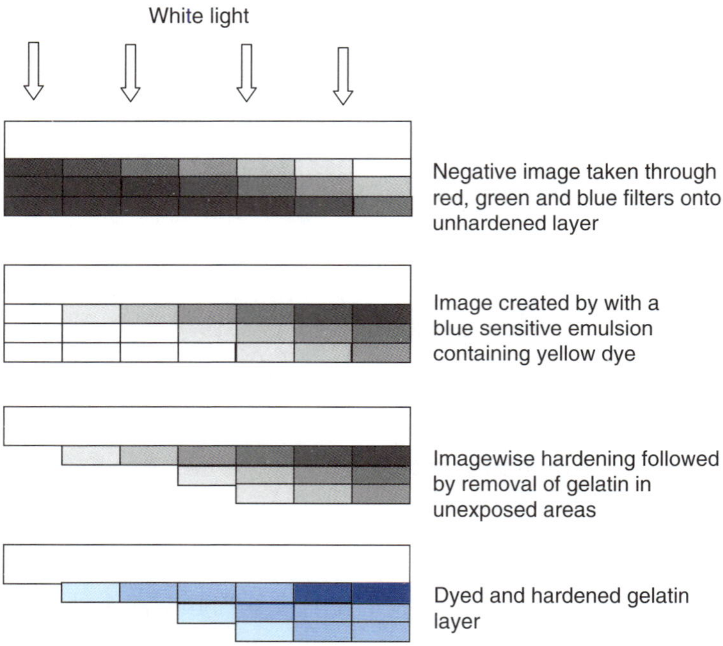

Figure 5 *The Kodak Dye Transfer Process*

developer scavenger) reduced. Under these circumstances, the oxidised pyrogallol reacts with gelatin, hardening the layer in the image area only.

This technique of partial layer hardening was not used often, being limited to some uses in the professional motion picture industry, and to a small extent in professional still photographs.

References

1. S.E. Sheppard and R.C. Houck, The structure of gelatin sols and gels I. The viscosity of gelatin solutions, *J. Phys. Chem.*, 1930, **34**, 273–298(Kodak Research Laboratories, Communication, No. 395).
2. S.E. Sheppard and J.G. McNally, The structure of gelatin sols and gels II. The anisotropy of gelatin gels, *Colloid Symposium Annual*, 1930, **7**, 17–39(Kodak Research Laboratories, Communication No. 399).
3. S.E. Sheppard and R.C. Houck, The structure of gelatin sols and gels III. Isoelectric points of gelatin, *J. Phys. Chem.*, 1930, **34**, 2187–2201(Kodak Research Laboratories, Communication No. 433).
4. F. Moll, H. Rosenkranz and W. Himmelmann, *Photographic Gelatin II*, R.J. Cox (ed), Academic Press, London, 1976, ISBN 0-12-194452-2, 197.
5. D.M. Burness and J. Pouradier, *The Theory of the Photographic Process*, 4th edn, T.H. James (ed), Maxmillan Publishing Co. Inc., New York, 1977, ISBN 0-02-360190-6, 77–87.
6. R.J. Cox (ed), *Photographic Gelatin II*, Academic Press, London, 1976, ISBN 0-12-194452-2, 143.
7. US 4,753,871, Eastman Kodak Co., K.N. Kilminster and D. Hoke.
8. US 4,775,616, Eastman Kodak Co., K.N. Kilminster and D. Hoke.
9. GB 1,255,787, Eastman Kodak Co., D. Macarthur Burness and C.J. Wright.

CHAPTER 3

Light Capture and Amplification

Silver halide crystals have been used as the light capturing component of the photographic process for well over 100 years. In the early years the only practical method of creating a coloured, faithful reproduction of an original scene was to handpaint photographs.

In the latter half of the nineteenth century (1886), however, Gabriel Lippmann proposed a general theory for his process of colour reproduction. He presented his work to the Paris Academy of Sciences in 1891, and in 1893 presented coloured photographs taken by A. and L. Lumière. This was followed by his complete theory of photography, published in 1894. The culmination of his photographic work was the award of the 1908 Nobel Prize for Physics.

The citation for his award read

for his method of reproducing colours photographically based on the phenomenon of interference

Many scientists and photographers contributed to the practice of photography. Lippmann is perhaps unusual in that an emulsion type is named after him. For example, two references to Lippmann emulsions appear in *The Theory of the Photographic Process.*

"Lippmann" emulsions have an average grain size of about 0.050 μm, and some coarse grain x-ray emulsions have crystals a few micrometers in diameter ....[1]

and

... the especially high contrast obtained by the development of Lippmann-type emulsions in developers such as Kodak developer D-8 is connected with their high concentration of potassium bromide, which promotes solution physical development of silver ions from neighboring grains onto the actively developing grains ....[2]

To this day silver halide crystals are precipitated, usually into a gelatin solution, during the reaction of alkali or ammonium halide with silver nitrate. There are many variables, which affect the size and shape of the resultant silver halide grains including

- gelatin concentration
- temperature
- concentrations of the halide and nitrate solutions
- sequence of addition
- rates of addition.

Modern emulsions are subject to precipitation using control of vAg or pH and therefore may not require a fixed volume of addition (Figure 1).

The silver and halide salt solutions are added to a location close to the stirrer paddle. The shape of the stirrer paddle and the speed of rotation are also factors in the eventual shape of the grain. Additionally the type of halide is also an issue. Silver halide emulsions can be pure silver chloride, bromide or iodide. They may also be mixtures of all or some of the infinite combinations of ratios between either two of the halides mentioned, or indeed all three.

Through rigorous investigations, all of the photographic manufacturers have been able to precipitate a range of sizes and shapes of silver halide crystals, depending upon the eventual application. The size of an emulsion grain is determined by standard optics. Simple microscopy can be used to furnish an image of the silver halide crystals, the angle of light and the shadow length of which will provide the size of the crystal, so long as a spherical bead of known diameter is also included in the

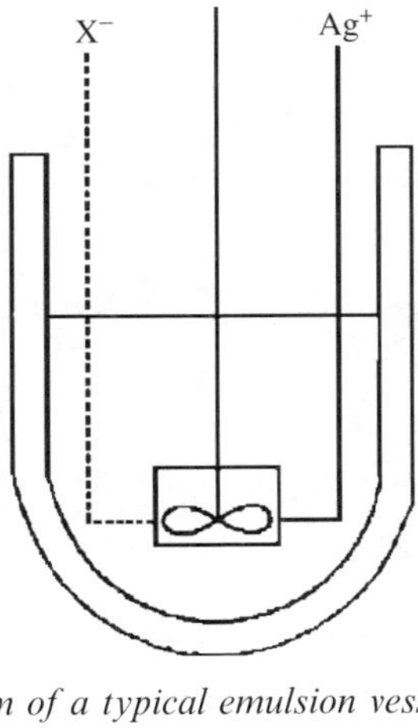

Figure 1 *A schematic diagram of a typical emulsion vessel*

micrograph. In this example, Figure 2, the sizing beads have been omitted.

If the silver halide crystals are spherical, no more data is needed, other than to determine the size range of the crystals. For even the very best of the emulsion, makers are unable to generate crystals that only have one size. The issue at hand is to minimise the range of sizes and any extraneous shapes that might be inadvertently precipitated.

Non-spherical emulsions will require additional data in order to determine the thickness of the grain. The closer the silver halide grains are to being flat, the more important grain thickness becomes, as this will determine the number of silver halide grains for a given silver halide laydown. A scanning electron micrograph will help to determine the thickness of the thin tabular grains, as the colours from the scanning electron micrograph are related to the thickness of the crystals. There are two distinct phases producing silver halide crystals, namely 'nucleation' and 'growth'. Nucleation is the term used to describe the creation of new crystals, and growth the addition of new material to the crystals that were precipitated in the nucleation step. Under certain circumstances, higher temperatures and in the presence of solvents, the smaller crystals may re-dissolve and the larger crystals grow still larger. This is known as 'Ostwald ripening'. There is another phenomenon, which may occur – that of 'recrystallisation'. This process involves crystals whose chemical composition is different. Under these circumstances, new

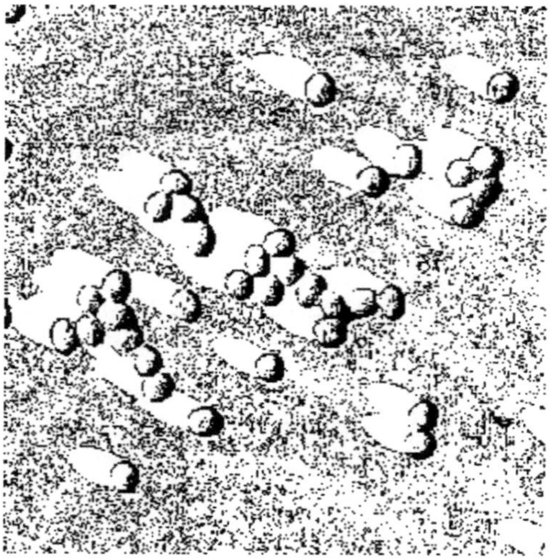

Figure 2 *A micrograph of silver halide crystals*

crystals are formed which are hybrids of the two chemically different types of silver halide crystals formed originally.

Of the many possible crystal shapes, some have been used in photographic products. Cubic silver halide grains were used for many years; latterly tabular silver halide grains have been used, certainly by the two largest photographic manufacturers. A representation of the tabular grain shape is shown in Figure 3.

It is also possible to generate extra facets, Figure 4.

More complex crystal shapes are possible and have been reported in the patent literature (see for example ref 3), which reported the following shape, Figure 5.

In their native state these crystals of any shape will have little sensitivity to light. The inherent sensitivity is for blue light, but this is small. The addition of a chemical sensitisation process at the end of the

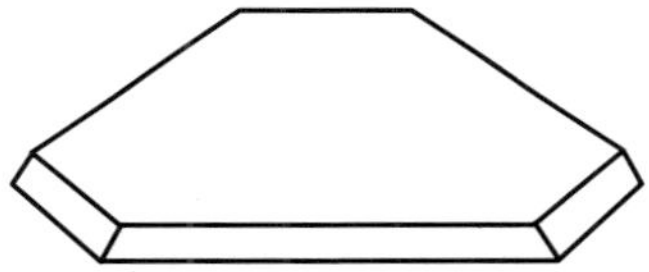

Figure 3 *Tabular shaped crystals*

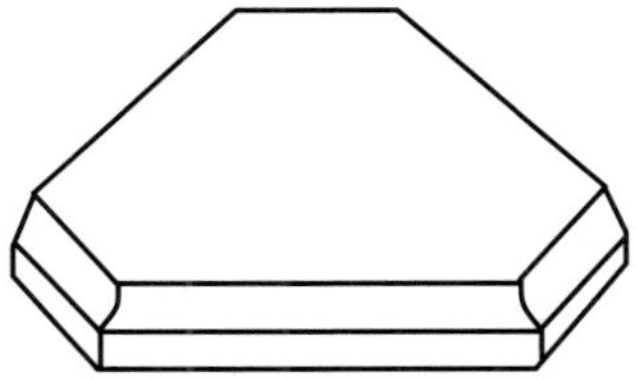

Figure 4 *An alternative tabular shape*

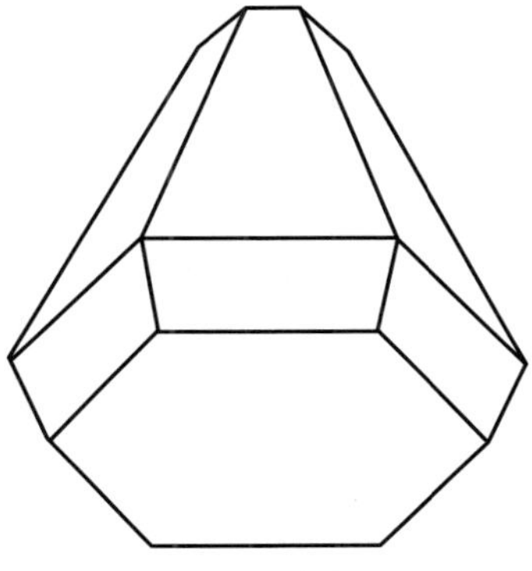

Figure 5 *Growth of a tabular grain on one of the grain surfaces*

emulsion growth sequence enables the silver halide crystals to readily capture most, if not all, of the available light. However, at this stage there is still no sensitivity to wavelengths of light outside the blue region. There are several types of chemical sensitisers, which include

- Sulfur/selenium
    - as thiosulfate salts
    - thioureas
    - etc.

- Gold or sometimes other heavy metals
    - as tetrachloroaurate (III)
    - as dithiocyanatoaurate (I)
    - potassium tetrachloroplatinate (II)
    - potassium tetrachloropalladate (II)
    - etc.

Practical chemical sensitisation may involve a mixture of, for example, sulfur and gold compounds. Chemical sensitisation will increase the photographic speed of the silver halide emulsions, hereafter referred to as 'emulsions'. A patent from the Fuji Photo Film Co. Ltd.[4] detailed several gold and silver salts which were of use at the time

- potassium tetra-chloroaurate
- auric trichloride (AuCl)
- potassium aurithiocyanate (KAu(CNS))
- potassium aurothiocyanate ($KAu(CNS)_2$).

The sulfur sensitising agents included

- sodium thiosulfate pentahydrate
- allyl thiocarbamate
- thiourea allylisothiocyanate
- ammonium and/or metal thiocyanate.

They also mention a reducing sensitisation agent, stannous chloride, which could be used in conjunction with the gold salt and also the use of poly-ethylene glycol.

In a further patent,[5] the Fuji Film scientists outline a simple formula for precipitating a light sensitive emulsion thus

Solution 1 (maintained at 48°C)	
Gelatin	*10 g*
Potassium iodide	*0.8 g*
Aqueous ammonia (10 N)	*18 ml*
Distilled water	*430 ml*
Solution 2	
A 0.01 wt. aqueous solution of potassium ferricyanide	*7 ml*
Solution 3 (maintained at 45°C)	
Silver nitrate	*80 g*
Distilled water	*300 ml*
Solution 4 (maintained at 48°C)	
Potassium bromide	*33.6 g*
Sodium chloride	*16.8 g*
Aqueous solution of ammonium rhodium chloride (0.01 by weight)	*0.5ml*
Distilled water	*350ml*

The procedure for the emulsion precipitation is as follows:

> *. . . While stirring Solution (1) in the dark, Solutions (2), (3) and (4) were added simultaneously over a period of 50 min. Immediately, sulphuric acid was added to adjust the pH of the mixture to 4.5. The mixture was allowed to stand for 20 minutes at 48°C. With further addition of 110 g of gelatin, the mixture was allowed to stand for 30 min. The resulting emulsion was cooled to gelation, shredded and washed with cold water for 1 h. The emulsion was heated to 60°C to re-dissolve the gelatin, and the pAg value adjusted to 7.2 by addition of an aqueous solution of cadmium chloride. With the further addition of 5 ml of 0.01 wt % aqueous gold tetrachloride the emulsion was left to stand for 60 min . . . .*

This emulsion formula is taken from an early patent (1971) and should be reasonably easy to reproduce in the modern laboratory with minimal equipment. Modern emulsion grain shapes are produced in highly sophisticated equipment with a high degree of process control so that pump rates, temperature and stirrer speeds are all maintained at a level of accuracy that is easy to reproduce.

Recent trends by most, if not all, photographic manufacturers have been to concentrate their efforts on the precipitation of tabular emulsions. Indeed in their 1989 patent,[6] R.H. Piggin, P.J. Zola and Ming J. Lin, comment

> *... The recent tabular grain emulsions have been observed to provide a large variety of photographic advantages, including, but not limited to*
>
> - *improved speed-granularity relationships*
> - *increased image sharpness*
> - *a capability for more rapid processing*
> - *increased covering power*
> - *reduced covering power loss at higher levels of forehardening*
> - *higher gamma for a given level of grain size dispersity*
> - *less image variance as a function of processing time and/or temperature variances*
> - *higher separations of blue and minus blue speeds*
> - *the capability of optimising light transmission or reflectance as a function of grain thickness*
> - *and reduced susceptibility to background radiation damage in very high speed emulsions.*
>
> *It has been recognised that still further improvements in emulsion sensitivity without any increase in granularity can be realised by forming recent tabular grain silver bromoiodide emulsions with iodide non-uniformly distributed within the grains....*

In the search for ever-faster emulsions which allow for the use of lower light levels or indeed product design trade-offs, the emulsion maker must be aware of the propensity of the emulsion to create fog centres, that is exposed silver where there was no incident light. This is of particular importance in colour paper products where there is a need to produce pure white in the non-exposed areas of a photographic print.

Compounds that act as anti-fogging agents have been patented in the literature for many years. These compounds will suppress the propensity of an emulsion to fogging, without the loss of desired photographic responses. Some of these materials are derived from benzotriazole and include

5-chlorobenzotriazole, 5-bromobenzotriazole, 5-methylbenzotriazole, 5-nitrobenzotriazole, 4,6-dinitrobenzotriazole, 1-ethylbenzotriazole and 1-phenylbenzotriazole.[7]

Anti-fogging compounds are not restricted to additives. The Fuji Photo Film Co. has been working on combining the desired features of an anti-fogging agent with its sensitising dyes, for example,[8] where they describe the synthesis of various analogues, Figure 6.

Anti-fogging compounds are used in most if not all colour emulsions as the most modern applications require that the photographic products be either stored or used in roll formats. Colour paper is stored in roll format during the manufacture, transport and pre-printing of customer prints. This requires a degree of flexibility in the coatings and therefore the emulsion grains, so that pressure-induced fog is kept to a minimum. The emulsion maker can, at least in part, ensure that the emulsion finish, *i.e.* the quantities of sulfur/gold and sensitising dyes are such that the finished emulsion is not on a knife edge for creating fog centres, yet not so slow (in photographic terms) that the emulsion cannot be used. Additionally some of the pressure effects can be addressed by ensuring that the silver halide grains in the coatings are kept at a safe distance from each other, which is part of the role of the product builder who is responsible for the design of the overall product.

For all practical applications, a photographic product needs to be able to capture light of longer wavelengths than is possible through natural sensitivity of the silver halide crystals. The sensitivity of a film to light of different wavelengths can be determined by exposing the material to light in a wedge spectrometer. These spectrometers typically contain a diffraction grating and a test wedge which will provide a spectrograph, Figure 7, which record the wavelengths of light to which this particular film is sensitive (from left to right). They also measure the sensitivity of this particular film to those wavelengths, by an inspection of the height

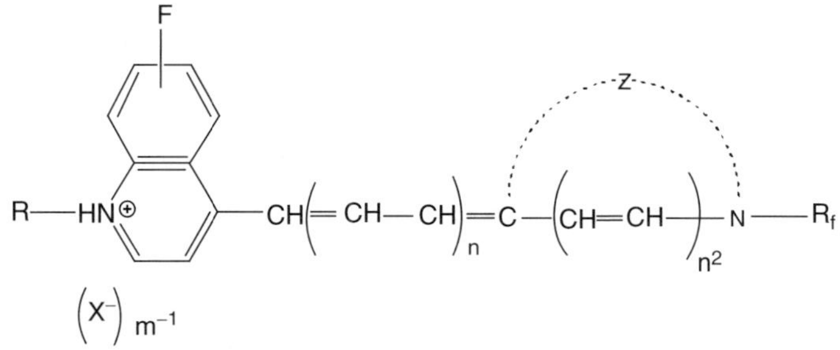

Figure 6 *The general structure for an antifogging agent with sensitising properties*

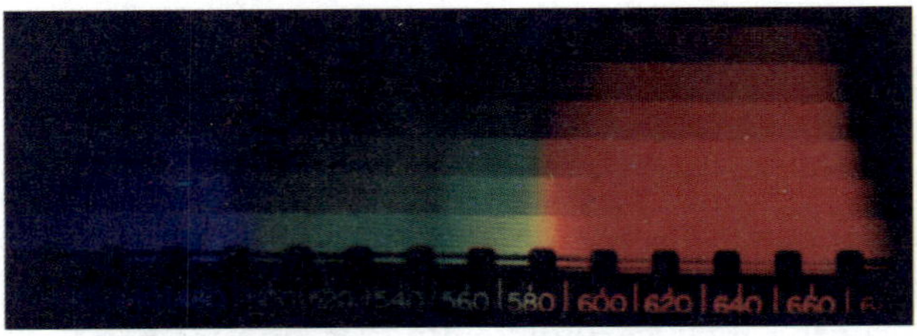

Figure 7 *A wedge spectrograph of a typical colour negative film*

or number of wedges of that wavelength. In this particular case the film has less sensitivity to blue light compared with red.

The method of achieving spectral adsorption was first discovered by Vogel[9,10] and was subsequently named spectral sensitisation. There are several basic rules of thumb, which were established by Eder.[11,12] Although these principles have been subsequently modified to a certain extent they are still in use today, and are

- to be effective, the dye has to be adsorbed onto the silver halide grain surface,
- the spectral characteristics of the adsorbed dye determines the spectral sensitisation of the silver halide grain, and
- the latent image from both intrinsic and spectral sensitisation is essentially the same.

The dye set that is to be used needs to have reasonable overlap, but not so much that there are unwanted exposures. Additionally the λ_{max} of the respective dyes should not leave a 'hole' in the visible spectrum. Half-band width is therefore of equal importance.

A specific property of these dyes is their ability to wash out during processing, otherwise they would leave an unacceptable stain in either the film or the paper. Dyes used for blue sensitisation require a λ_{max} of 465–485 nm, green λ_{max} of 500–600 nm and red between λ_{max} 600–700 nm. The exact value of the peak depends upon the application (paper, film, *etc.*) as well as the type of film (amateur, professional, infrared for example). In some cases the desired adsorption may only be achieved by using a blend of two dyes per colour record. There is a need to consider dye aggregation and adsorption onto the grain surface, which is a topic that appears in the major texts – see for example ref 13. Order of addition is important when using dye mixtures. If two dyes are to be adsorbed onto one grain surface one of the dyes will adhere more than the other. Also some dyes aggregate together at certain concentrations, which may lead to other issues. The most effective dyes currently used for spectral sensitisation are the cyanine and merocyanine dyes. Other dye classes, including acridine orange, alizarine blue, congo red, eosin, erythrosine and ethyl violet, were evaluated with varying degrees of success in the early years (Figure 8).

The accreditation and scale-up process of any new material is such that changes take place slowly, no more so than with sensitising dyes. Arguably, much of the ground breaking research concerning cyanine dyes was undertaken in the 1940s, with patents for most of the useful dyes being filed in the 1970s. Although many patents have been filed since that time concerning new structures for sensitising dyes, only a very few have made the transition from laboratory scale novel compounds to industrial scale use in products. Some of the more modern patents provide the reader with an interesting glimpse of more recent developments in sensitising dye thinking – even if the patents did not result in the use of the reported dyes in products. Where relevant, some of these patents will be mentioned.

The colour of the dye is related to the length of the conjugation. Additionally, there is a nomenclature within each of the three dye families mentioned above. An example from the cyanine family is presented below with the λ_{max} of the dye, Figure 9.

Cyanine dyes can be prepared *via* a number of different pathways. A generic synthesis is provided in the following scheme, Figure 10.

Much of the synthetic routes to cyanine dyes published by Eastman Kodak Co. were developed by Brooker and his co-workers in the 1940s

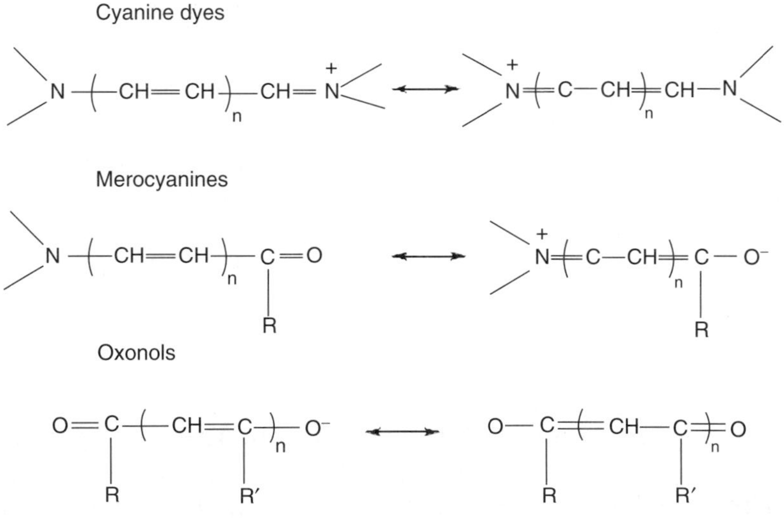

Figure 8 *Generic formulae for the three classes of dye*

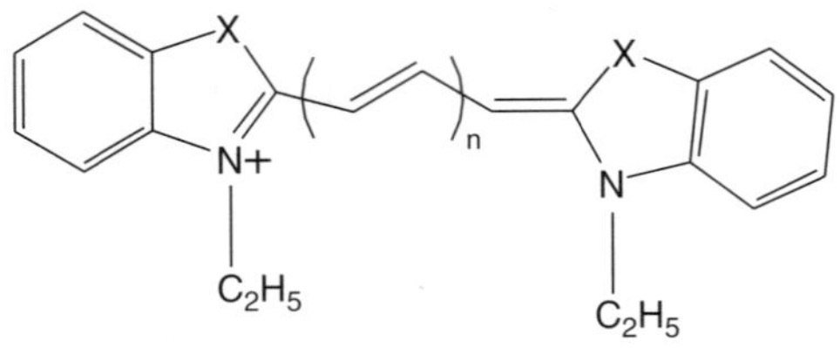

		λ_{max}[a]
n=0	a simplecyanine	423
n=1	a carbocyanine	557
n=2	a dicarbocyanine	650
n=3	a tricarbocyanine	758

[a]methanolic or ethanolic solutions

Figure 9 *The effect of increasing the conjugation on the λ_{max} of the dye*

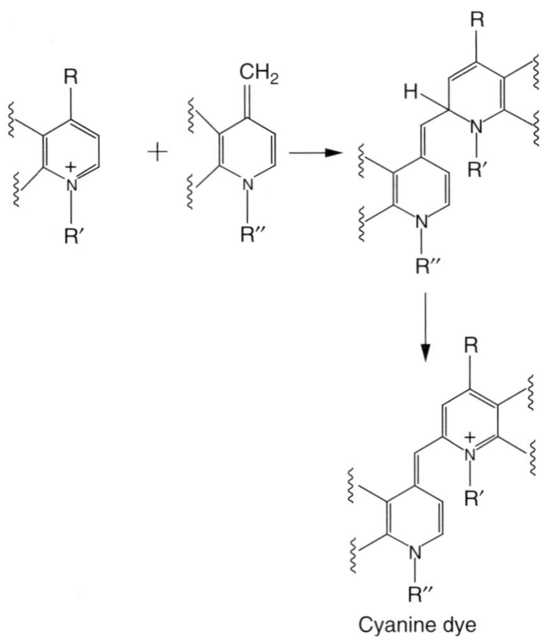

Figure 10 *Generic synthesis of cyanine dyes*

and 1950s, see ref 14 and 15 for further details. These are by no means the only route to cyanine dyes, but they were the basis of the later Eastman Kodak Co. patents regarding synthetic pathways. Figure 11 details a practical route to cyanine dyes reported by Brooker.[14]

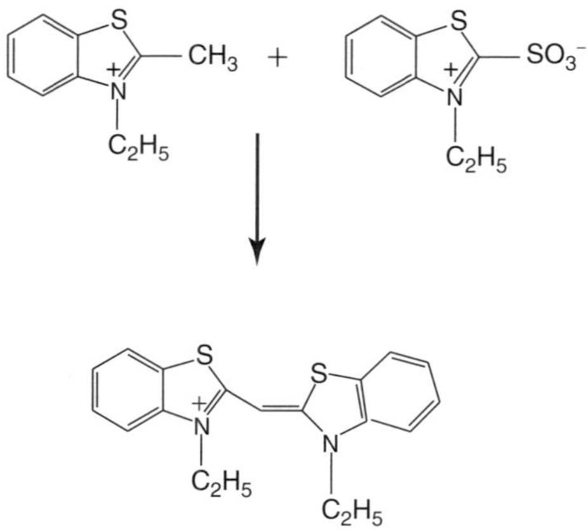

Figure 11 *Brooker's route to a cyanine dye*

3.1 Blue Sensitisation

Not only is the photographic application important, *i.e.* film, paper or transparency, the type of silver halide grains to which the sensitising dye is to be adsorbed is also of importance. For example, colour negative films tend to be manufactured using emulsions that contain silver bromoiodide crystals. This grain type/application tends to use one of the following two dye families in layers that are designed to capture information for the blue record of the image, Figure 12.

Emulsions used in paper products tend to be silver chloride or silver chlorobromide. These emulsions are often sensitised using merocyanine dyes. One example is provided as Figure 13.

Or they use derivatives of rhodanine, Figure 14.

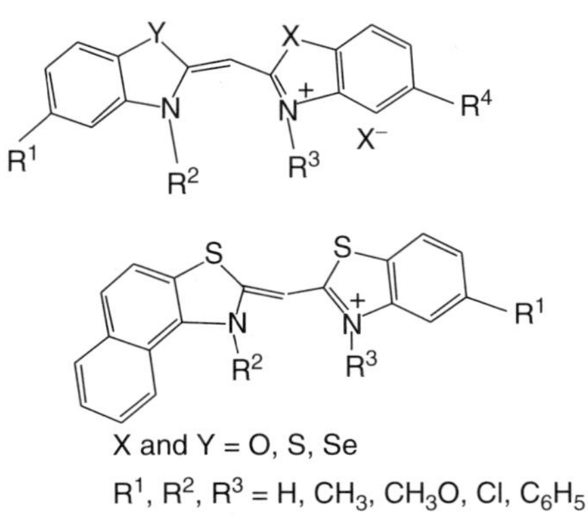

Figure 12 *Examples of blue sensitising dyes for colour negative films*

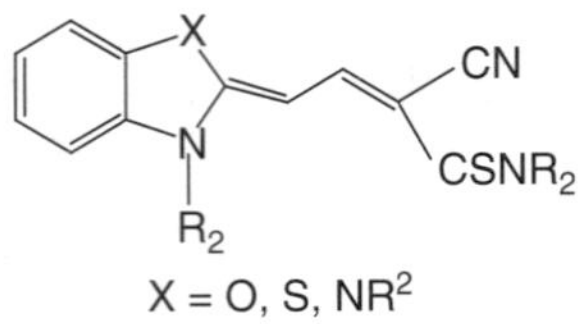

Figure 13 *A blue sensitising dye used in colour paper*

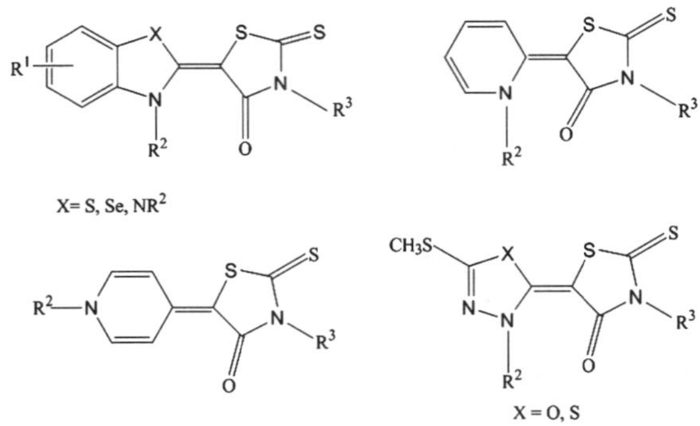

Figure 14 *Some rhodanine blue sensitising dyes*

3.2 Green Sensitisation

The sensitisation for green and red light records is somewhat more complex, as the various films and papers tend to use two sensitising dyes in combination, each of different λ_{max} values. There is now a further complexity in that the ratio between the dyes may vary. Furthermore, different silver halide compositions, *i.e.* silver bromide, iodide, bromo-iodide or chloroiodide may require different dye sets. A sample of the cyanine dyes that have been used in products appears in Figure 15.

Merocyanine dyes that have been used in colour films, papers, medical X-rays, direct positive films and black and white papers are detailed in Figure 16.

3.3 Red Sensitisation

Sensitising dyes for the red region of the spectrum is application specific. Figure 17 shows the λ_{max} values that are needed for the various products.

Red sensitising dyes also tend to be used in pairs. A typical pair of dyes used in colour film applications is given below, Figure 18.

A longer wavelength pair of film red sensitising dyes appears as shown in Figure 19.

This particular combination of dyes has also been used in some black and white paper applications. These dyes have also been used in

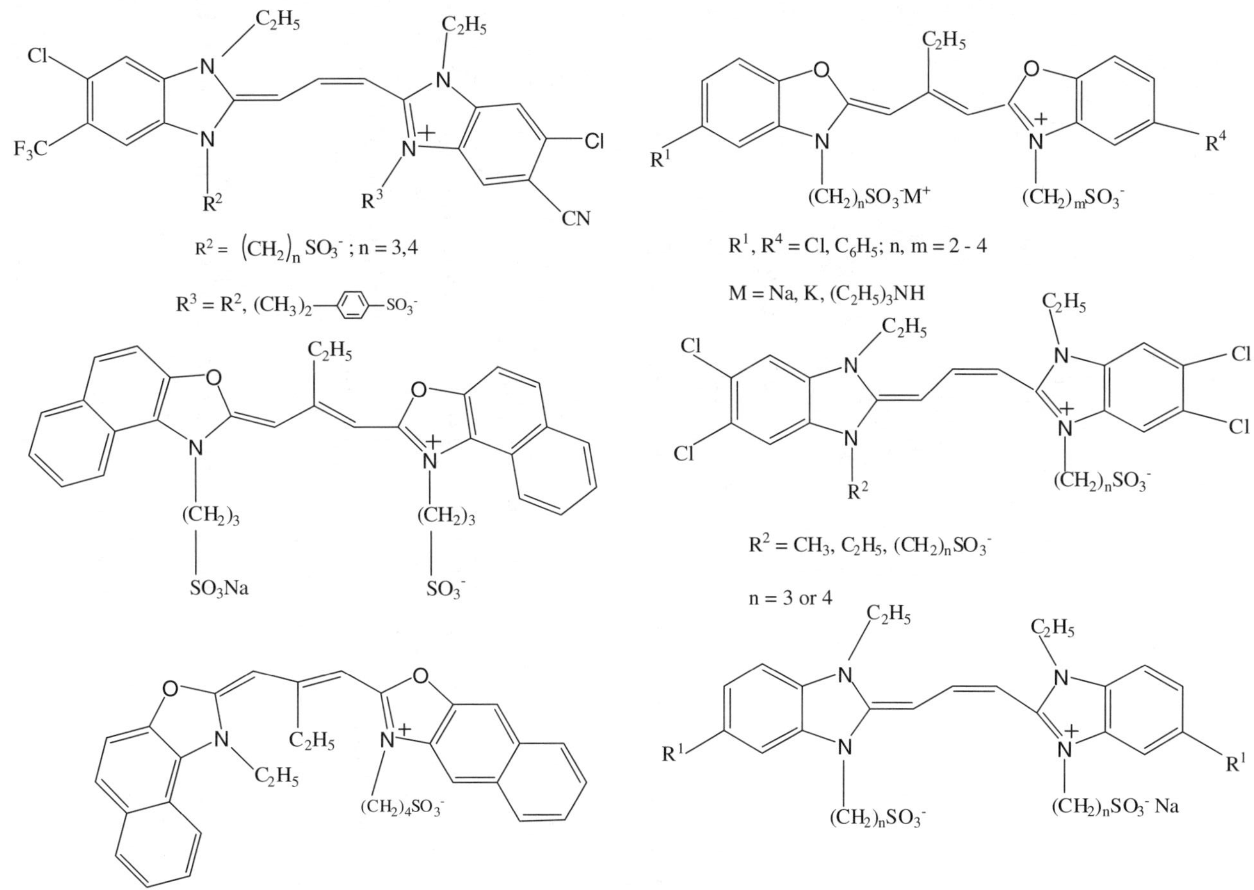

Figure 15 *Examples of green cyanine sensitising dyes*

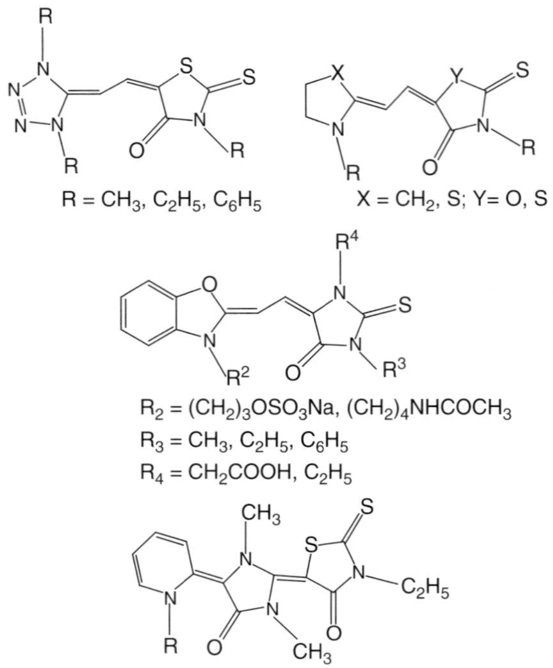

Figure 16 *Examples of merocyanine green sensitising dyes*

Application	λmax in the red region
Colour negative films, colour reversal films, colour reversal papers	600-660nm
Colour negative papers, pan chromatic black and white papers	Maximum sensitivity at or above λ_{max} 700nm
Panchromatic black and white films	Up to λ_{max} 650nm
Aerial photographic films	Panchromatic sensitisation (up to λ_{max} 750nm)

Figure 17 λ_{max} *values for red sensitising dyes*

conjunction with a third dye, Figure 20, in aerial survey film, where the spectral region of the aerial films also covers the near infrared.

Some more recent alternatives for use as infrared sensitising dyes have been published by Agfa-Gevaert,[16] Figure 21.

where

- R1 and R4 each independently represent a hydrogen, an alkyl, an aryl or a sulfonic acid group,
- R2 and R3 each independently represent a hydrogen or an alkyl group,

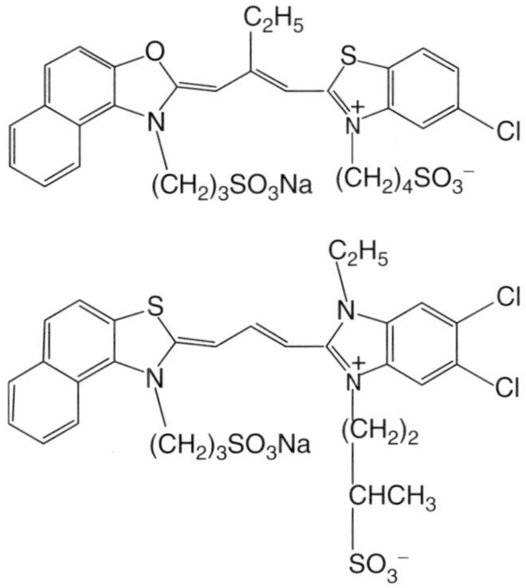

Figure 18 *Typical dyes used for red sensitisation in colour negative films*

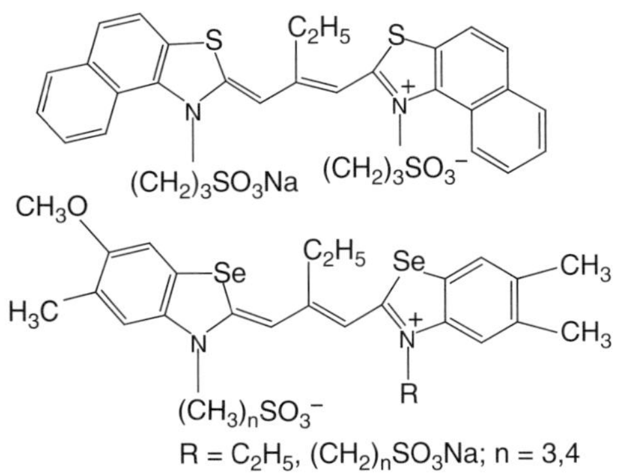

Figure 19 *Longer wavelength red sensitising dyes*

- R5 and R6 each independently represent a hydrogen, or sulfonic acid group,
- *n* and *m* each independently represents an integer from 1 to 6,
- and at least R1 and R4 or R5 and R6 represent a sulfonic acid group.

The sensitising dyes discussed thus far have concentrated on colour applications. The graphic arts industry uses many types of films and some paper products in the preparation of printed material. These products used a range of exposure devices with different light sources. Clearly each laser wavelength used as a light source would require a sensitising dye commensurate with that wavelength. This is a subject in

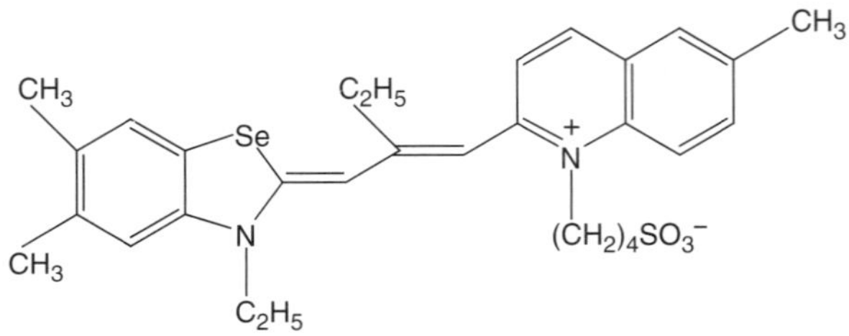

Figure 20 *A near infrared sensitising dye used in aerial films*

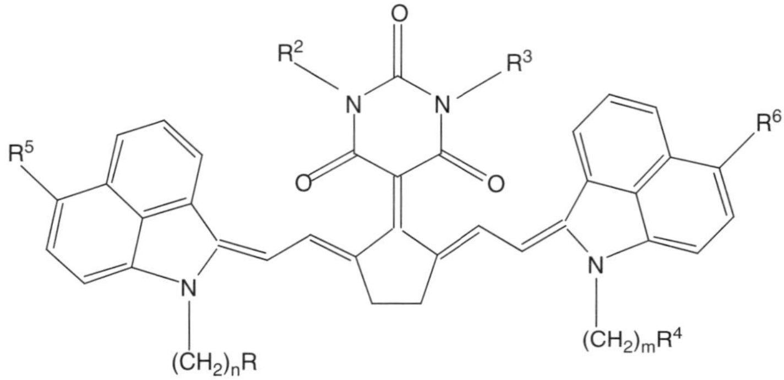

Figure 21 *An infrared sensitising dye*

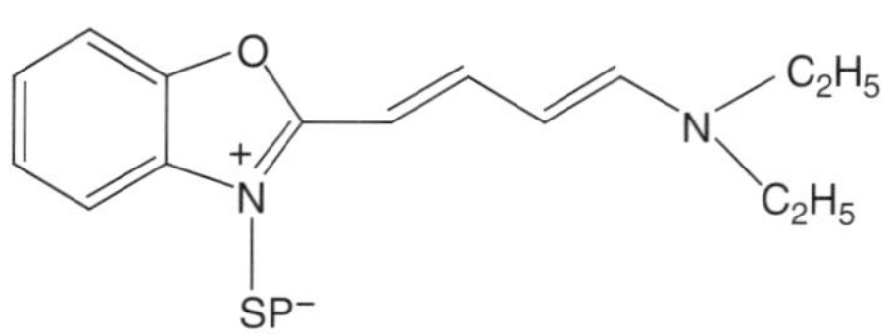

Figure 22 *A typical sensitising dye used in graphic arts films*

its own right. Perhaps by way of example, the types of compounds disclosed in[17] might serve to illustrate the types of chemistry use in graphic arts products and to note the similarities/contrast and the differences between the colour sensitising dyes. The dye family is represented by Figure 22.

The patent describes the uses of this dye family thus

> ... *The silver halide used in the practice of the invention can be of any known type, such as silver bromoiodide, silver bromide, silver chloride, silver chlorobromide and the like. The silver halide grains may be of any known type, such as spherical, cubic or tabular grains. In a preferred embodiment,*

the dyes of the invention are used to sensitise a cubic black and white graphic arts emulsion.
The amount of sensitising dye that is useful in the invention is preferably in the range of 0.1 to 1.0 mmol per mole of silver halide and more preferably from 0.2 to 0.7mmol per mole of silver halide. Optimum dye concentrations will depend on the intended end use of the photographic material and can be determined by methods known in the art.
The silver halide to be used in the invention may be subjected to chemical sensitisation with compounds such as gold sensitisers (e.g., aurous sulfide) and others known in the art ....

Sensitising dyes used in black and white products have been the subject of many patents. Konishiroku Photo Industry Co. Ltd. published a patent in 1971,[18] which detailed some of the sensitising dyes that they investigated for their black and white products. This particular dye, which can also be used in colour photographic applications, was designed for silver bromoiodide emulsions containing 3 mole % of silver iodide, Figure 23.

Whatever the photographic product, the first stage of producing an image is that of light capture by the silver halide crystal. The image formed by this exposure is called a latent image. A more detailed discussion of the amplification and development of the latent image into the final image will be discussed later. In concept the process is outlined in Figure 24.

The quantity of incident light or other incident radiation is usually such that there is no analytical procedure that can be used to detect a difference in the silver halide crystal. The developer solution contains a

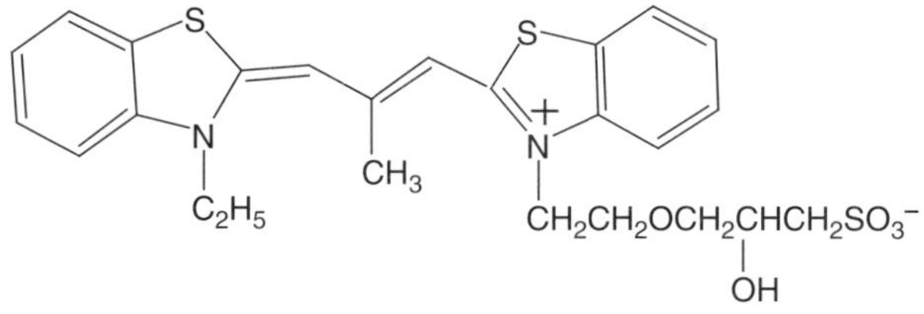

Figure 23 *A typical black and white sensitising dye*

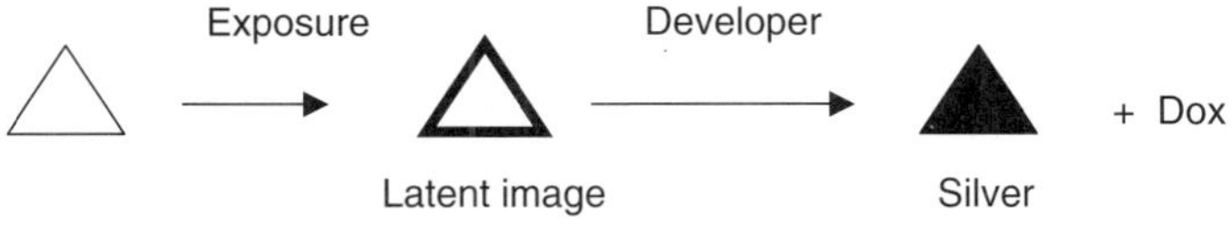

Figure 24 *The conversion of silver ions to silver during development*

reducing agent that causes an exposure dependent amount of the silver halide to be converted to silver metal.

References

1. T.H. James (ed), *The Theory of the Photographic Process*, 4th edn, Macmillan Publishing Co. Inc., New York, 1977, ISBN 0-02-360190-6, 100.
2. T.H. James (ed), *The Theory of the Photographic Process*, 4th edn, Macmillan Publishing Co. Inc., New York, 1977, ISBN 0-02-360190-6, 418.
3. M. Saitou, S. Aiba, S. Yamada, E. Okutsu, US 2003/0224305, Fuji Photo Film Co. Ltd.
4. GB 1,298,254, Fuji Photo Film Co. Ltd.
5. GB 1,244,818, Fuji Photo Film Co. Ltd.
6. R.H. Piggin, P.J. Zola and M.J. Lin, US 5,061,616, Eastman Kodak Co.
7. GB 1,298,119, Fuji Photo Film Co. Ltd.
8. GB 1,210,009, Fuji Photo Film Co. Ltd.
9. H.W. Vogel, *Chem. Ber.*, 1873, 6, 1302.
10. H.W. Vogel, *Chem. Ber.*, 1876, 9, 667.
11. J.M. Eder, *Photogr. Corresp.*, 1885, 22, 349.
12. J.M. Eder, *Photogr. Corresp.*, 1886, **23**, 146.
13. T.H. James (ed), *The Theory of the Photographic Process*, 4th edn, Macmillan Publishing Co. Inc., New York, 1977, ISBN 0-02-360190-6, 251–290.
14. L.G.S. Brooker, R.H. Sprague, C.P. Smyth and G.L. Lewis, *J. Amer. Chem. Soc.*, 1940, **62**, 1116.
15. L.G.S. Brooker, G.H. Keyes, R.H. Sprague, R.H. Van Dyke, E. van Lare, G. van Zandt, F.L. White, H.W.J. Cressman and S.G. Dent Jr., *J. Amer. Chem. Soc.*, 1951, **73**, 5332.
16. EP 1, 093, 015, Agfa-Gevaert, G. Deroover.
17. EP 0,512, 483, Eastman Kodak Co., J. D. Mee.
18. GB 1,253,839, Konishiroku Photo Industry Co. Ltd.

CHAPTER 4

Developers

Many compounds, both organic and inorganic, have been evaluated as potential developing agents, more commonly called developers. The basic property of a developer is that the molecule has to be a reducing agent. Not all reducing agents, however, are developers, as they need to be able to differentiate between exposed and unexposed silver halide crystals. While having the ability of discrimination as a basic requirement, there are further properties that are essential including

- the ability to form dyes of the relevant hues when the oxidised colour developer reacts with the coupler,
- stability of the developer solution to temperature, pH, oxygen, *etc.* relevant kinetic properties, *i.e.* activity/kinetics
- *etc.*

The older texts, see for example *Photographic Processing Chemistry*,[1] often compile a list of organic compounds that have been evaluated for use as developers. Figure 1 provides a representative sample, rather than an exhaustive list, of those different compound types which have been investigated and shown to be developers.

All of the photographic manufacturers market their own, unique, developer solutions. For each product family, however the developing agent is common. For example, a colour negative film developer solution from all of the photographic manufacturers will contain the same chemical, known in literature as CD4. This is because all of the manufacturers want as much of the market for their own solutions as well as their films. Precluding other manufacturer's films from using their own solutions limits the potential market. Thus developers are industry-standard. The developer solutions, from which the developer is made, will vary as each manufacturer tries to emphasise a particular quality, such as throughput rates, stability, *etc.* Figure 2 outlines a brief list of common developers and their applications.

Name	Compound	Date	Originator
1,2,3-trihydroxybenzene (pyrogallol)	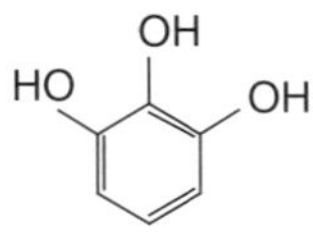	1850	Archer
1,2 dihydroxybenzene (catechol)	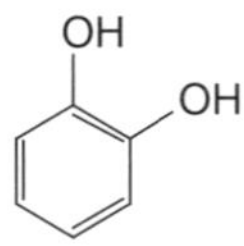	1880	Eder and Toth
1,4-dihydroxybenzene (and derivatives)	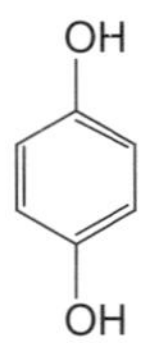	1880	Abney
1,4-diaminobenzene hydrochloride (and derivatives)	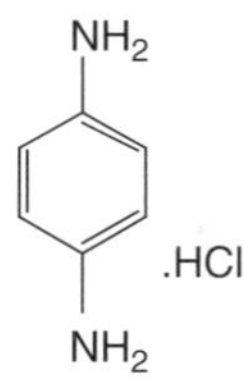	1888	Andresen
1,2,4-triaminobenzene	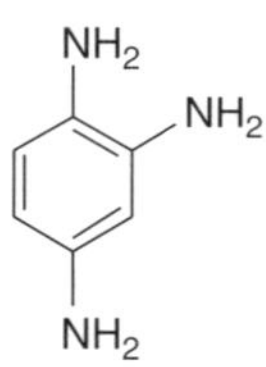	1907	König and Staehlin
Gallic acid esters	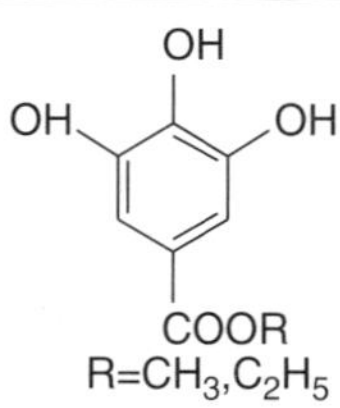	1908	Menter
1,2-dihydroxynaphthalene	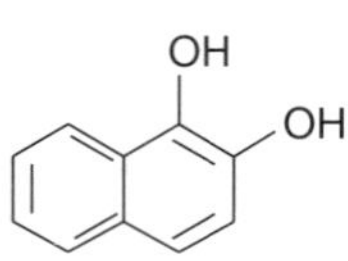	1930	Andresen and Leupold

Figure 1 *Some common developing agents*

Name	Compound	Date	Originator
Ascorbic acid	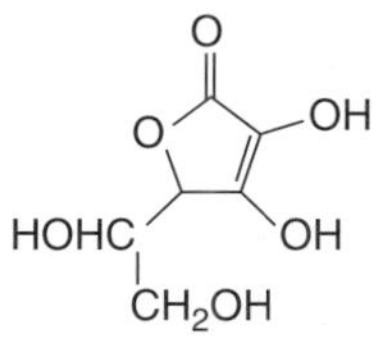	1932	Öhle
2-oxytetronic acid (and derivatives)	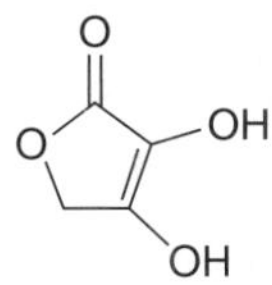	1935	Maurer and Zapf
N-(3-methyl-4-hydroxyphenyl) tetrahydropyrrole	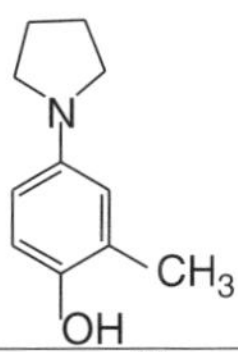	1962	Mason, Gauguin, Ramsay and Kaye
Phenidone Z	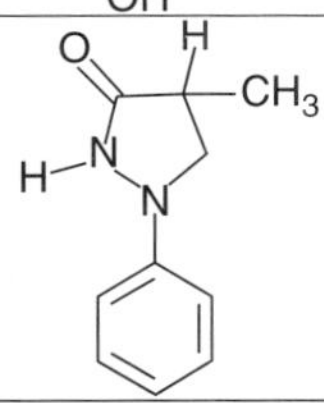		Ilford Ltd
Dimezone	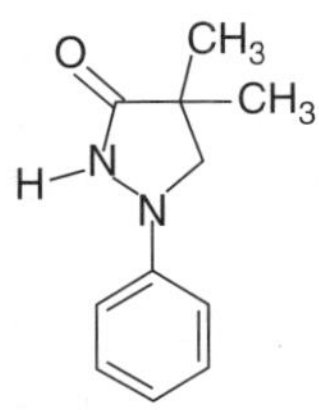		Eastman Kodak Co.

Figure 1 (*Continued*)

4.1 Black and White Developers

Various black and white films are still commercially available. These films are processed through a number of developer solutions that are known as D-76, HC-110, DK-50 and D-19 – at least in the Eastman Kodak Co. literature. These developers are made up in solutions which contain a number of other chemicals that are added to control pH and aerial oxidation, for example. These black and white developer solutions are sold either as concentrates, that require dilution with water, or more

Name	Structure	Product
Hydroquinone	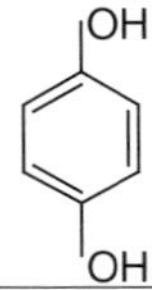	black and white films graphic arts films
4- (N-ethyl-N-2-hydroxyethyl) -2-methylphenylenediamine sulphate	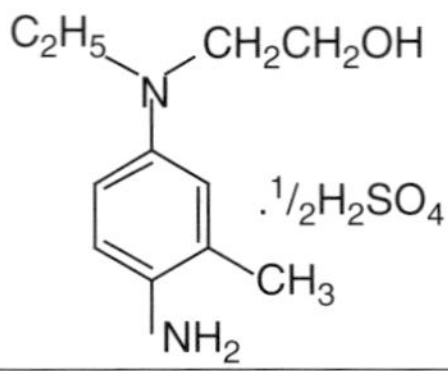	colour film
4- (N-ethyl-N-2-methanesulphonylaminoethyl) -2-methylphenylenediamine sesquisulphate monohydrate	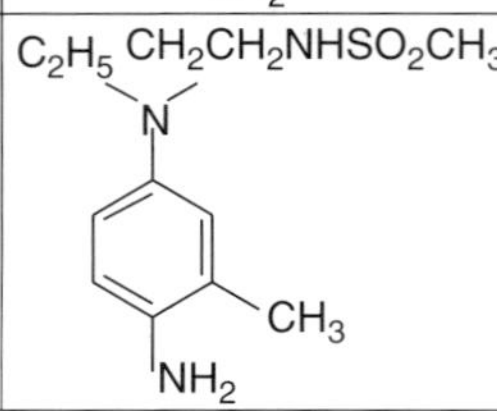	motion picture, colour paper

Figure 2 *Some commercial developing agents*

than one concentrated solution, which may require dilution after the two solutions are mixed together. Chapter 5 considers the chemical formulae of the four Kodak developer solutions mentioned above. Inspecting the formulae shows that all four contain hydroquinone as the developing agent. Indeed hydroquinone is used in the majority of commercial black and white developing solutions.

The issue with black and white development is not that of producing a silver image, so much as the 'fate' of the oxidised developer. Hydroquinone is oxidised to benzoquinone during silver development thus, Figure 3.

The fate of the benzoquinone is of paramount interest, as left in solution it may cause stain or undergo a number of reactions, see for example Figure 4.

Developer solutions are therefore designed to minimise these side reactions, details for which appear in Chapter 5.

For completeness, graphic arts films also use hydroquinone for developing the latent image. More film format details are given in Chapter 9. Motion picture films use the same *p*-phenylenediamine as colour paper developer. The film structure and more details of the unique chemistry of motion picture films appear in Chapter 12.

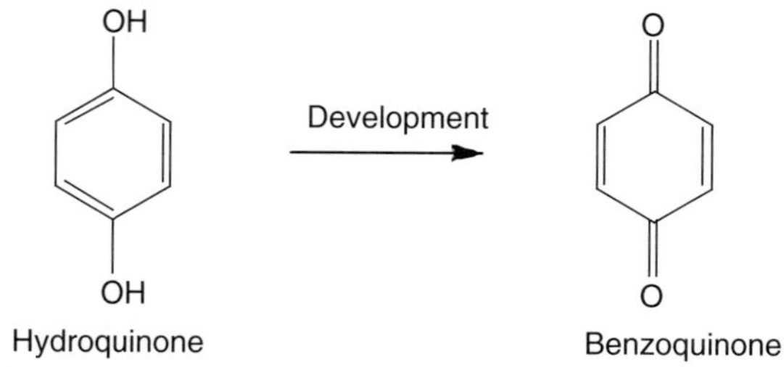

Figure 3 *Benzoquinone is formed during development*

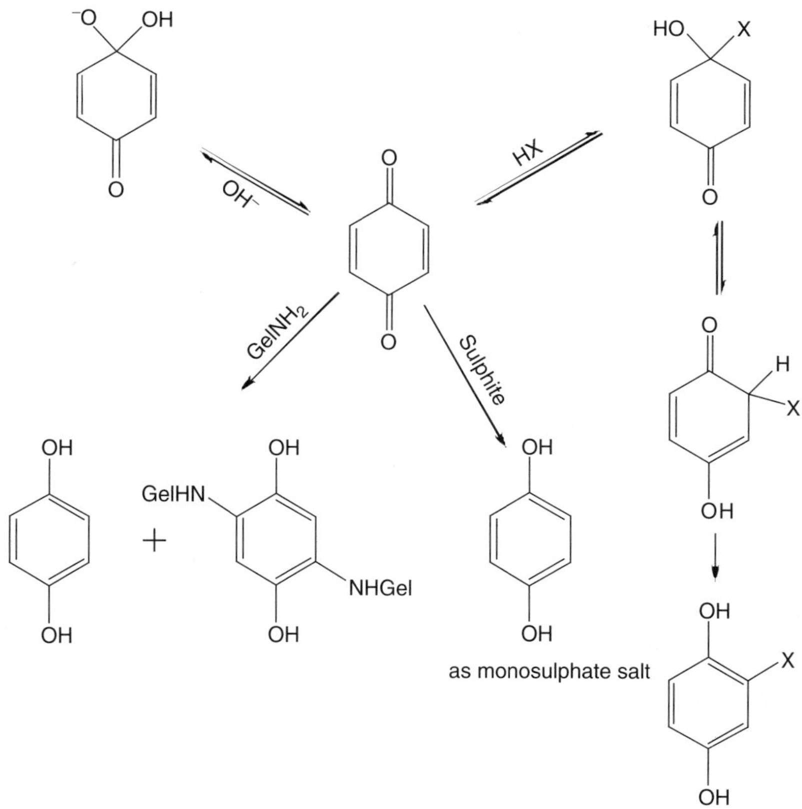

Figure 4 *Some of the many side reactions of benzoquinone*

4.2 Colour Films and Paper Developers

The most important colour developers, derivatives of which are used in both colour negative film and colour paper, are *p*-phenylenediamines.

Eastman Kodak Co. introduced both the film and paper developer formulations, which are now the industry-standards. They are known as the C41 process for colour negative film, and the RA-4 for colour paper.

The dyes needed for colour film and colour paper are different because the application is different. In colour film the processed negatives are often stored in paper or cardboard wallets, where glue from the wallets can react with the dyes. Humidity during storage may also be an issue. They are also designed for use with transmitted light.

While there are chemicals that could be potential developers, *p*-phenylenediamines have been proven to be the most useful for producing dyes. These developer molecules must donate an electron to the silver ion during the process of becoming oxidised. An alkyl group *ortho* to the coupling amino group has been found to facilitate this electron donation, thereby increasing the developer activity. Most of these *p*-phenylenediamine compounds cause dermatitis, the sensitivity to which increases with prolonged exposure. These materials are therefore treated with caution, particularly during the manufacturing process.

The synthesis of many colour *p*-phenylenediamine developers is described by Bent *et al.*[2] in their 1951 paper. Of the 54 developers or so that are described two of them are used in products. They are known as CD4, used as the industry-standard for processing colour negative film, and CD3, used as the industry-standard for processing colour paper. CD4 was first described a patent from a German company now known as Agfa-Gavaert.[3,4] The synthesis of CD4, as described by Bent *et al.*, is given in Figure 5.

The colour paper developer, CD3, is also used in motion picture film and was patented first by Weissberger.[5] Bent *et al.* report that the synthesis of this compound is through the nitroso intermediate outlined in Figure 5. In his patent, Weissberger *et al.* claimed that substitution into the aryl ring afforded dyes (on oxidation and subsequent reaction with couplers) whose hue could be changed. They commented:[5]

> ... these compounds may be substituted in the aromatic ring with other groups including ... they have a tendency to alter the colour of the final dye image and the colour may be controlled in this way ...

The resultant dye hue, or colour, is but one of a number of tests that are listed in the paper published by Bent *et al.* They carried out tests on each of the 54 or so *p*-phenylenediamine developers that they report, namely

- half wave potential
- development rates

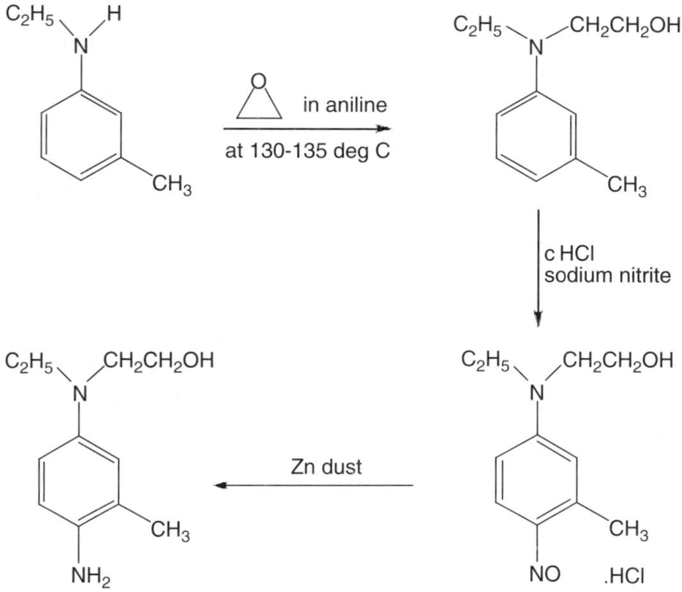

Figure 5 *The synthesis of CD4*

- coupling efficiencies
- biological assays.

The half wave potential provides a measure of the free energy change of the oxidation potential of the developer. A higher free energy for a given compound is indicative of the tendency of the developer to release an electron, which is a mandatory process for a developer. Using the Lewis–Randall convention,[6] a good developer will exhibit a more positive half wave potential compared with a poor developer. Figure 6 compares CD3 and CD4 with the *N*,*N*-diethyl parent developer.

These results suggest that there is no difference between the three compounds in terms of their ability to release electrons.

Development rates for the three compounds were measured in coatings as the time taken to attain an optical density of 0.2 above fog (the density produced from an unexposed emulsion), using a light source with an exposure of logE 1.75. The measurements (as 1/time, units of min^{-1}) for the three developers listed in the order shown in Figure 6 were 0.80, 0.38 and 0.67, respectively, suggesting that the *N*,*N*-diethyl developer is more efficient in terms of development rates.

Coupling efficiencies were measured using a standard coupler, in this case 2-cyanoacetyl-coumarone. The technique used was to evaluate the eluted dye after chromatographic separation of the mixture produced upon reaction between the coupler and oxidised colour developer. All three of the developers listed in Figure 6 showed comparable coupling efficiencies.

The above evidence would suggest that all three developers could be used for any application, provided that the dye hue and dye stability were acceptable for the given application. The determining factor was the biological assay tests. The basic test was to evaluate the developers for skin sensitisation in guinea pigs. CD3 and CD4 proved to be less dermatitic than did the parent compound listed in Figure 6. This test is of paramount importance because the developer solutions are handled in areas where air extraction may not be possible, and potentially in large vats for the larger photofinishing companies.

Earlier on in this chapter the development rate of CD3 was reported as being considerably lower than either CD4 or the parent *N*,*N*′-diethyl

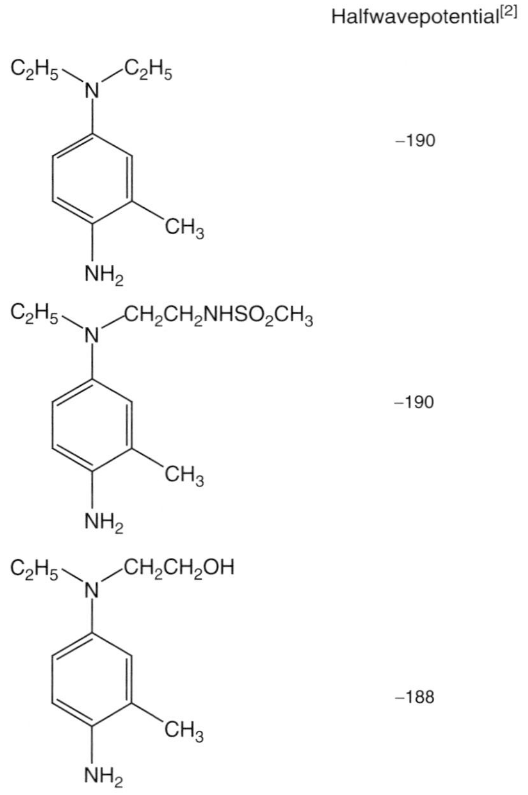

Figure 6 *A comparison of colour developers*

p-phenylenediamine. The developer solution made using CD3 for colour paper applications has been enhanced by the addition of benzyl alcohol – see Chapter 5 for further details. Indeed developers are used in conjunction with other chemicals to form a developer solution, which is only one step in the development of a colour negative film or paper. Figure 7 outlines the use of the various solutions in some of the commercial processing kits that are available.

Figure 8 shows the effect on the film components when the film is processed through the various solutions. The chemical composition of the various solutions is covered in Chapter 5.

In broad outline, a latent image is formed on exposure to light. During colour development the silver halide crystals that have been exposed to light are converted to silver, at the same time that the image dye is formed from the oxidised colour developer and the relevant coupler. The bleach step converts the silver back to silver ions, and the fix step removes the silver ions. Unreacted coupler molecules that remain in the film are harmless as they are not coloured.

Figure 8 is not intended to be to scale, nor does the diagram contain all of the layers that would be found in any colour product, for example

Colour Negative Film	Colour Prints from Colour Negatives	Colour Reversal/ Chrome Film	Black-and-White Negatives	Black-and-White Prints
process C-41	process RA-4	process E-6	—	—
developer	developer	first developer	developer	RC: developer
bleach	bleach-fix	wash	stop	stop
wash	stabiliser or wash	reversal bath	fix	fix
fixer	dry	colour developer	rinse	wash
wash		pre-bleach	washing aid	fibre: developer
stabiliser or final rinse		bleach	wash	stop
dry		fixer	drying aid	fix
		wash		rinse
		final rinse		washing aid
		wash		wash
		dry		

Figure 7 *Processing steps for several product families*

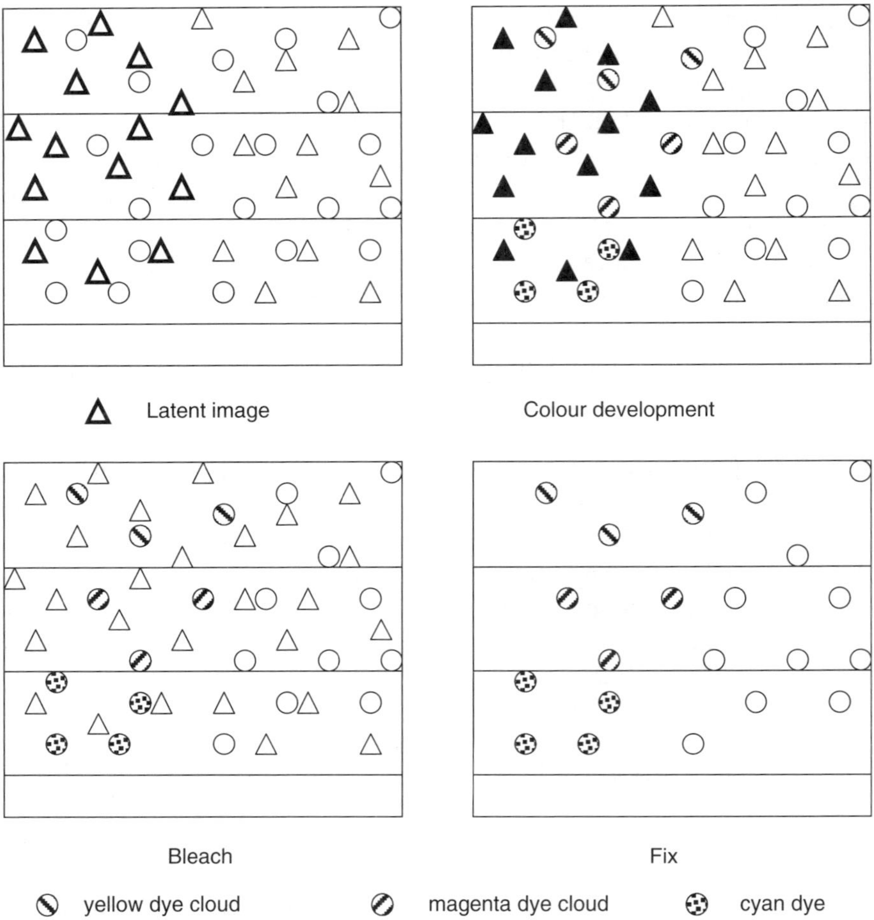

Figure 8 *Chemical changes during the development steps*

the yellow filter layer has been omitted. Chapters 9–12 cover film and paper design in much more detail.

The chemical formulations of some of the commercially available processing solutions are covered in Chapter 5.

References

1. L.F.A. Mason, *Photographic Processing Chemistry*, Focal Press Ltd., London, 1975, ISBN 0-240-50824-6.
2. R.L. Bent, J.C. Dessloch, F.C. Duennebier, D.W. Fassett, D.B. Glass, T.H. James, D.B. Julian, W.R. Ruby, J.M. Snell, J.H. Sterner, J.R. Thirtle, P.W. Vittum and A. Weissberger, *J. Amer. Chem. Soc.*, 1951, **73**, 3100.

3. US 2,108,243 Agfa-Gavaert, B. Wendt.
4. GB 460,580, Agfa-Gavaert, W.W. Groves.
5. US 2,193,015, Eastman Kodak Co., A. Weissberger.
6. G.N. Lewis and M. Randall, *Thermodynamics*, McGraw Hill Inc., New York, 1923.

CHAPTER 5

Processing Solutions

All commercially available developer solutions are buffered to pH values commensurate with their uses. Although many compounds are capable of forming buffers, only a few are of use in photographic systems. Figure 1 provides some of the alternative buffering systems and their pH ranges which have been tested with photographic materials. This list is not intended as exhaustive, but indicates the range of chemicals, and their respective pH values, available to the processing chemists who formulate these solutions.

A stable developer pH range is not the only criterion for a commercial developer solution. Water purity, or more accurately water hardness, caused by soluble salts is another issue. Sequestering agents are added to some of the developer formulae to remove unwanted ions. These sequestering agents tend to be derivatives of ethylenediaminetetraacetic acid (EDTA) and are very efficient at forming complexes with metal ions such as copper and iron. Various metal ions such as Fe^{2+}, Ti^{3+} and V^{2+} are capable of reducing silver ions to silver. For this reason and others, free ions must be removed from the solutions.

Developer solutions are often exposed to the air for long periods of time. Under these circumstances the developer may become aerially oxidised which can result in bi-products that produce a brown stain. A

Buffer	pH range
$K_2SO_3/K_2S_2O_5$	6.5-8.0
$Na_2SO_3/Na_2S_2O_5$	6.5-8.0
$Na_2B_40_7/HBO_3$	8.0-9.2
$Na_2CO_3/NaHCO_3$	9.0-11.0
$Na_2B_4O_7/NaOH$	9.2-11.0
KBO_2/KOH	11.0-12.0
$NaBO_2/NaOH$	11.0-12.0
$Na_3PO_4/NaOH$	12.0-13.0

Figure 1 *Some photographically tested buffers*

sulfite salt is therefore added to most developer solutions so that any developer that becomes aerially oxidised is removed by the sulfite anions. Although sulfite anions are but one type of preservative, they are the most commonly used in photographic processing solutions.

In some circumstances, unexposed silver halide grains may develop in the absence of light. This phenomenon is known as fog. Bromide ions are more commonly added as anti-foggants, usually as potassium salts, although organic anti-foggants may also be used.

Some developer solutions, notably those used to process colour paper, contain benzyl alcohol as an accelerating agent for the dye-forming reaction. Thiocyanate ions may also be used as accelerating agents.

Formulations of various solutions are presented below, which demonstrate the use of many of these compounds, either in isolation or combination. The precise formulae of the solutions is application-specific, as they are tailored to meet the needs of the chemicals in the various films and papers. There is a need for all photographic manufacturers to ensure that their particular products can be used with any other manufacturer's films and papers. While this need maximises the potential market for each product, the downside is that the basic chemistry in these solutions has remained constant for many years. In the case of colour film development, for at least the last 20 years.

There are many different commercially available processing solutions for all of the various product types. Figure 2 details those processing solutions that are used for most of the more common photographic products. For comparative purposes, the recommended processing solution for three Fuji Film products is given alongside the equivalent Kodak product.

Chapter 4 discussed the synthesis or structure of a number of developers. Listed below are the processing solutions for a representative sample of photographic products.

Product type	Processing solution – Eastman Kodak Product name	Processing solution – Fuji Film Product Name
Colour Film	C-41	CN-16, CN-16Q, CN-16FA, CN-16L, CN-16S or C-41
Colour Paper	RA-4	RA-4
Motion Picture Film	ECN-2	ECN-2
Black and white Film	D76, HC-110, DK-50, D-19	
Black and white paper	RA-4	
Transparency (Ektachrome)	E-6	
Reversal Paper (Radiance)	R-3	
Transparency (Kodachrome)	K14M	
Graphic Arts Film	RA-2000	

Figure 2 *Commercially available processing solutions*

5.1 Black and White Processing Solutions

Processing black and white films and papers is less complex than the colour materials. The basic steps (see aj3 on the Kodak website[1]) are

- film developer
- stop bath (or water)
- fixer
- hypo clearing agent (optional).

Development is terminated with the stop bath, and the remaining silver removed using the fixer. A stop bath is recommended, as it extends the life of the fixer solution.

Each developer solution produces different results (see document o3 – Kodak website[1]) assuming that one were to process identical rolls of film through the various developer solutions. The aforementioned Kodak document o3, provides the following comparison

D-76	for general use
HC-110	produces results similar to those produced by KODAK developer D-76
DK-50	provides normal contrast, average to slightly higher than average graininess
D-19	provides higher than normal contrast and speed, higher than average graininess.

The consumer therefore has a choice of how the film is processed and also on what paper to print the image. For a discussion of the paper choices – see the Kodak website (document o3). The description of the different modes of manufacture, for example fibre-based, resin coated, graded papers and variable contrast papers are also given a brief explanation.

The developer solutions are

D-76 (Figure 3 a and b), HC-110 (Figure 4 a and b), DK-50 (Figure 5 a–c), D-19 (Figure 6 a and b).

A typical stop bath, in this case KODAK Royal Print Stop bath consists of water and acetic acid as shown in Figure 7(a). For a typical concentrated fixer solution (for KODAK POLYMAX T Fixer), see Figure 7(b) and for the corresponding working strength solution, see Figure 7(c).

Figure 8 details the recommended development times (see o3 from the Kodak website) for various films.

D-76

Concentrate

Weight %	Component
85-90	Sodium sulphite
1-5	Hydroquinone
1-5	Sodium tetraborate
1-5	Bis(4-hydroxy-N-methylanilinium) sulphate
< 1	Boric anhydride
< 1	Pentasodium (carboxylatomethyl)iminobis(ethylenenitrilo)tetraacetate

(a)

Working solution: (approximate dilution - 110 g concentrate to make 1 litre)

Weight %	Component
85-90	Water
5-10	Sodium sulphite
< 1	Sodium tetraborate
< 1	Hydroquinone
< 1	Bis(4-hydroxy-N-methylanilinium) sulphate
< 0.1	Boric anhydride
< 0.1	Pentasodium (carboxylatomethyl)iminobis(ethylenenitrilo)tetraacetate

(b)

Figure 3 *D-76 developer solutions*

HC-110

Concentrate

Weight %	Component
65-70	Water
15-20	Diethanolamine-sulphur dioxide complex
1-5	Hydroquinone
1-5	Diethylene glycol
1-5	Ethanolamine
≤ 1	Diethanolamine
≤ 1	Ethylene glycol

(a)

Working solution: (approximate dilution - 100 ml concentrate to make 1 litre)

Weight %	Component
90-95	Water
1-5	Diethanolamine-sulphur dioxide complex
< 1	Hydroquinone
< 1	Diethylene glycol
< 1	Ethanolamine
< 1	Diethanolamine
< 1	Ethylene glycol

(b)

Figure 4 *HC-110 developer solutions*

DK-50

Part A

Weight %	Component
45-50	Hydroquinone
45-50	Bis(4-hydroxy-N-methylanilinium) sulphate

(a)

Part B

Weight %	Component
75-80	Sodium sulphite
15-20	Sodium metaborate
1-5	Potassium bromide

(b)

Working solution: (approximate dilution - 5 g A + 37.9 g B to make 1 litre)

Weight %	Component
95-100	Water
1-5	Sodium sulphite
< 1	Sodium metaborate
< 1	Hydroquinone
< 1	Bis(4-hydroxy-N-methylanilinium) sulphate

(c)

Figure 5 *DK-50 developer solutions*

D-19

Concentrate

Weight %	Component
55-60	Sodium sulphite
25-30	Sodium carbonate
10-15	Hydroquinone
1-5	Bis(4-hydroxy-N-methylanilinium) sulphate
1-5	Sodium hexametaphosphate
1-5	Potassium bromide

(a)

Working solution: (approximate dilution - 160 g concentrate to make 1 litre)

Weight %	Component
85-90	Water
5-10	Sodium sulphite
1-5	Sodium carbonate
< 1	Hydroquinone
< 1	Bis(4-hydroxy-N-methylanilinium) sulphate
< 1	Sodium hexametaphosphate
< 1	Potassium bromide

(b)

Figure 6 *D-19 developer solutions*

In Chapter 4 (Figure 2), the use of hydroquinone as a developer was mentioned for both black and white and graphic arts products. Even though the basic developer is identical, the developer solution used for graphic arts processing, marketed as RA-2000 by the Eastman Kodak Co., is completely different to the black and white processing solution

A typical stop bath, in this case KODAK Royal Print Stop bath consists of:

Weight %	Component
80-85	Water
15-20	Acetic acid

(a)

A typical concentrated fixer solution would be (for KODAK POLYMER T Fixer)

Weight %	Component
50-55	Water
30-35	Ammonium thiosulphate
1-5	Sodium acetate
1-5	Sodium bisulphite
1-5	Ammonium sulphite
1-5	Acetic acid
1-5	Boric acid

(b)

And the corresponding working strength solution:

Weight %	Component
85-90	Water
10-15	Ammonium thiosulphate
1-5	Sodium acetate
1-5	Sodium bisulphite
1-5	Ammoniu sulphite
< 1	Acetic acid
< 1	Boric acid

(c)

Figure 7 *Stop and fixer solutions for KODAK POLYMAX T*

formulae provided above. The chemistry for the production of half tones used for producing images in graphic arts products (discussed in Chapter 1, Figures 18, 19 and 20) will be discussed in Chapter 9. The chemistry for the developer solution is provided below, Figure 9.

Figures 3, 4, 5 and 6 show alternative developer solution formulations for processing either black and white films or graphic arts films. For comparative purposes, all the working strength formulations are presented in Figure 10, where the units for the quantities are weight per cent.

The interesting issue here is that the developer in each developer solution is hydroquinone, yet the supporting chemistry has been modified/designed for different products. The necessary film characteristics are the driver. Some films require higher contrast, some better control of fog, *etc*. More details concerning film and paper characteristics are given in Chapters 9–12.

5.2 Colour Film Processing Solutions

Chapter 4 considered the synthesis of various colour developers, each of which is a *p*-phenylediamine. Listed below in Figures 11, 12, 13 and 14

Development Times for Modified Developer D-76 in Rack-and-Tank Processors at 68°F (20°C)			
KODAK Film	Exposure	Optimum Development Time	Batched Development Time
T-MAX 100 Professional	Normal Push1 Push 2	8.0 9.0 11.5	8.0 10.0 12.0
T-MAX 400 Professional	Normal Push 1 Push 2	7.5 8.5 9.5	8.0 8.0 10.0
T-MAX P3200 Professional	EI 800 EI 1600 EI 3200 EI 6400	10.0 11.0 13.0 16.0	10.0 12.0 14.0 16.0
PLUS-X Pan	Normal Push 1 Push 2	6.0 7.0 10.0	6.0 8.0 10.0
TRI-X Pan	Normal Push 1 Push 2	7.5 8.5 9.5	8.0 8.0 10.0
Other Black-and-Film	Exposure	Optimum Development Time	Batched Development Time
AGFAPAN APX25	Normal Push1 Push 2	5.5	6.0
AGFAPAN APX 100	Normal Push 1 Push 2	6.5 7.5 10.0	6.0 8.0 10.0
AGFAPAN AP400	Normal Push1 Push 2	8.0 9.0 11.0	8.0 10.0 12.0
ILFORD Pan F	Normal	7.5	8.0
ILFORD FP4 Plus	Normal Push 1 Push 2	6.0 7.0 10.0	6.0 8.0 10.0
ILFORD HP5 Plus	Normal Push 1 Push 2	6.0 7.0 9.5	6.0 8.0 10.0
ILFORD 400 DELTA	Normal Push 1 Push 2	6.0 7.0 10.0	6.0 8.0 10.0
ILFORD 100 DELTA	Normal Push 1 Push 2	5.5 7.0 9.0	6.0 8.0 10.0
FUJI NEOPAN 400	Normal Push1 Push 2	7.5 9.5 11.5	8.0 10.0 12.0
FUJI NEOPAN 1600	EI 1600	7.5	8.0

Figure 8 *Black and white development times for various products*

Weight %	Component
60-65	Water
10-51	Potassium sulphite
5-10	Diethyleneglycol
5-10	Hydroquinone
4-10	Potassium carbonate
1-5	Sodium sulphite

Figure 9 *A typical graphic arts developer solution*

Component	D-76	HC-110	DK-50	D-19	RA-2000
Water	85-90	90-95	95-100	85-90	60-65
Potassium sulphite					10-15
Diethylene glycol		<1			5-10
Hydroquinone	<1	<1	<1	<1	5-10
Potassium carbonate					4-10
Sodium sulphite	5-10		1-5	5-10	1-5
Sodium tetraborate	<1				
Bis(4-hydroxy-N-methylanilinium) sulphate	<1		<1	<1	
Boric anhydride	<0.1				
Pentasodium (carboxylatomethyl)iminobis(ethylenenitrilo)tetraacetate	<0.1				
Diethanolamine-sulphur dioxide complex		1-5			
Ethanolamine		<1			
Diethanolamine		<1			
Ethylene glycol		<1			
Sodium metaborate			<1		
Sodium carbonate				1-5	
Sodium hexametaphosphate				<1	
Potassium bromide				<1	

Figure 10 *A comparison of various developer solutions*

are the processing chemicals, with their amounts, used for processing a typical professional colour film from the Eastman Kodak Co.

An example of a typical bleach solution, fix solution and final rinse solution (used with professional colour film) can be seen in Figure 12, 13 and 14, respectively.

Figure 15 compares the components from a selected Fuji Film developer solution with the C-41 process from the Eastman Kodak Co.

The RA-4 colour paper processing solutions are available in several types of kit. For example, the commercial kit for mini-labs is different from that used by the large photo finisher. There is a simple processing kit available that comprises of three separate solutions, which make up the developer solution and two that make up the bleach-fix solution. The unusual feature of the RA-4 process is that it combines the bleach and

Part	Component	Weight %
1		
	Water	80-85
	4- (N-ethyl-N-2-hydroxyethyl) -2-methylphenylenediamine sulphate	15-20
	Sodium bisulphite	<1
2		
	Water	90-95
	Diethylene glycol	5-10
	Mixture of C12-15 alcohol ethoxylates	1-5
	A mixture of: 5-chloro-2-methyl-4-isothiazolin-3-one and 2-methyl-4-isothiazolin-3-one	<0.25
3		
	Water	50-55
	Potassium carbonate	35-40
	Potassium sulphite	1-5
	Pentasodium (carboxylatomethyl) iminobis (ethylenenitrilo) tetraacetate	1-5
	Sodium sulphite	1-5
4		
	Water	70-75
	Bis (hydroxylammonium) sulphate	25-30

Figure 11 *A typical colour negative developer kit*

An example of a typical bleach solution, also for use with professional colour film.

Component	Weight %
Water	65-70
Succinic acid	5-10
Ferric ammonium propylenediaminetetraacetic acid	5-10
Ammonium bromide	5-10
Ammonium nitrate	1-5
Trimethylenediaminetetraacetic acid	1-5

Figure 12 *A typical bleach solution for colour film*

An example of a typical fix solution, also for use with professional colour film.

Component	Weight %
Water	65-70
Ammonium thiosulphate	25-30
Ammonium thiocyanate	1-5
Ammonium sulphite	1-5
Sodium sulphite	1-5

Figure 13 *An example of a fix solution for colour film*

An example of a typical final rinse solution (used with professional colour film)

Component	Weight %
Water	90-95
Diethylene glycol	5-10
Mixture of C12-15 alcohol ethoxylates	1-5
Mixture of 5-chloro-2-methyl-2H-isothiazol-3-one and 2-methyl-2H-isothiazol-3-one (3:1)	<1

Figure 14 *A typical final rinse solution for colour film*

Components – C41 Process
4- (N-ethyl-N-2-hydroxyethyl) -2-methylphenylenediamine sulphate
A mixture of: 5-chloro-2-methyl-4-isothiazolin-3-one and 2-methyl-4-isothiazolin-3-one
Bis (hydroxylammonium) sulphate
Diethylene glycol
Mixture of C12-15 alcohol ethoxylates
Pentasodium (carboxylatomethyl) iminobis (ethylenenitrilo) tetraacetate
Potassium carbonate
Potassium sulphite
Sodium bisulphate
Sodium sulphite
Water

Components for CN-16FA
Diethanesulfonic Acid
Diethylene Glycol
DTPA
Hydroxylamine Sulphate
Potassium Carbonate
p-Phenylenediamine
Sodium Sulphite
Water

Figure 15 *Eastman Kodak Co and Fuji Photo Film developer solution formulations each using the same developer*

fix processing steps. The actual formulation for this kit can be seen in Figure 16(a–c). This makes up a working developer solution of (using a dilution of 90 ml A+20 ml B+90 ml C to make 1 l), Figure 16(d).

Chapter 4 discussed the synthesis of 4- (*N*-ethyl-*N*-2-methanesulfonyl-aminoethyl)-2-methylphenylenediamine sesquisulfate monohydrate, which is the developer that is also used in the development of motion picture films. See Figure 17 for details of the working concentration of the developer solution.

Just as with the comparison of hydroquinone developer used in graphic arts and black and white film products, the formulations for the developer solutions made from 4- (*N*-ethyl-*N*-2-methanesulfonyl-aminoethyl)-2-methylphenylenediamine sesquisulfate monohydrate for RA-4 and motion picture films is completely different. This once again reflects the differences in the chemistry of the film and paper products that are to be processed through the various solutions, and the desired final photographic properties.

All of the processing solutions mentioned in this chapter will produce faithful reproductions of the original scene, be they negative or positive

RA-4 Developer - Part A

Component	Weight %
Water	70-80
Triethanolamine	10-15
N,N-diethylhydroxylamine	1-5
Substituted stilbene	1-5

(a)

RA-4 Developer - Part B

Component	Weight %
Water	60-70
4- (N-ethyl-N-2-methanesulphonylaminoethyl) -2-methylphenylenediamine sesquisulphate monohydrate	20-25
Lithium sulphate	10-15
Potassium sulphite	1-5

(b)

RA-4 Developer - Part C

Component	Weight %
Water	70-80
Potassium carbonate	20-25
Potassium bicarbonate	1-5
Potassium chloride	1-5

(c)

This makes up a working developer solution of (using a dilution of 90ml A + 20ml B + 90ml C to make 1 litre)

Component	Weight %
Water	90-95
Potassium carbonate	1-5
Triethanolamine	1-5
4- (N-ethyl-N-2-methanesulphonylaminoethyl) - 2-methylphenylenediamine sesquisulphate monohydrate	1
N,N-diethylhydroxylamine	1

(d)

Figure 16 *RA-4 developer solution chemistry*

Component	Weight %
Water	95-99
Sodium carbonate	1-5
4- (N-ethyl-N-2-methanesulphonylaminoethyl) - 2 -methylphenylenediamine sesquisulphate monohydrate	< 1
Substituted phosphonate	< 1
Sodium sulphite	< 1

Figure 17 *Working concentration for the RA-4 developer solution*

Figure 18 *The RA-4 control strip*

images, provided that the physical parameters that affect the processing solutions are kept in control. These control strips are available as unprocessed strips from the photographic manufacturers. The RA-4 process is typically used by a mini-lab operator or photofinisher who will process one of the strips, at an appropriate interval, through their process. A schematic diagram of an RA-4 control strip is shown below as Figure 18.

The optical densities of these patches are read on a densitometer and plotted on a control chart. Eastman Kodak Co. publication z130_07 (available from their website) details the procedure for care and control of the unexposed control strips and of plotting the final density values. If the suggested plots are within the control limits, the processing solutions are deemed to be in control and the process can be used for processing customer orders. If there is a problem, appropriate action needs to be undertaken prior to using the solutions.

Replenishing solutions are commercially available that allow the stock solutions to be topped up to the required concentrations. The above is one example of the use of control strips to monitor and control processing solutions. This system of monitoring the activity of the solutions is available for all of the commercially available processing kits. The objects on the control strip vary commensurate with the application of the film or paper.

Reference

1. www.kodak.com

CHAPTER 6

Colour Forming Couplers

A coupler is a chemical which when reacted with oxidised colour developer, produces a dye. Hundreds of thousands of chemicals have been screened as potential couplers but few have been or are used in commercially available colour films or papers. Although couplers can be added to the film layers without any pre-treatment, in practice most couplers are dissolved in a solvent and dispersed in gelatin. This material is referred to as a coupler dispersion or simply a dispersion. Chapter 7 contains more details of the process and the solvents that are used.

In screening chemicals the following criteria must be addressed

- the reaction rate with oxidised colour developer,
- the need to form colourless dispersions (in most cases),
- solubility of the coupler in the dispersion-making process,
- the need for the coupler and resultant dye to remain in a fixed location,
- the colour of the resultant dye, and
- stability of the resultant dye.

Couplers are designed such that part of the molecule contains a long side chain which prevents the coupler from moving during the coating and processing phases. These long side chains are referred to as 'ballasts', and are discussed later in this chapter.

Additionally, couplers are application-specific. The decision to use a particular coupler may depend on the exposure conditions of the film or paper, *i.e.* UV, daylight, *etc.* as well as its use in a reflection (paper) or a transmission (film) product. Chapter 4 outlined the use of different developers, and Chapter 7 discusses the dye formation and chemistry. Those chapters do not consider the effect that reflected light may have on the perceived dye density. Consider the situation of a single layer of cyan dye coated on a reflective surface, compared with the same dye coated on a transparent surface, Figure 1.

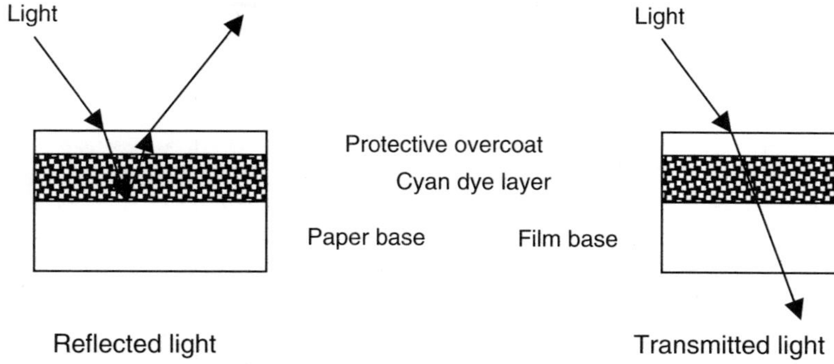

Figure 1 *Reflected v transmitted light*

In the case of reflection, the photons of light pass through the cyan layer twice, whereas in the case of transmission the photons will only pass through the layer once.

One might therefore expect that a plot of reflection density *vs.* transmission density would produce a straight line with a slope of 2. In practice there are slight losses, for example the paper base is not a perfect light reflector. Figure 2 shows a plot of reflected *vs.* transmitted light.

Additionally, all dyes have unwanted absorptions (see Chapter 8 which considers the chemistry of colour). These unwanted absorptions are of lower density than the wanted absorptions. In the case of reflected light these unwanted absorptions increase in density more than the wanted absorptions. The net effect is that the reflection dye curve appears broader than the corresponding transmission curve, once the differences in dye density have been taken into account, Figure 3. Hunt provides the explanation thus.[1]

> *... The three curves in figure 3 are for the same dye, but coated in a transparency format (t), a reflection format (p). Curve t' is the data from curve t multiplied by 2.61 for all spectral regions. This ensures that the λ_{max} values for the two curves therefore taking into account the transmission/reflection argument discussed above and shown in figure 2. In all cases curve p is broader with more unwanted absorptions (shaded area).*

6.1 Dye Formation

The overall reaction of a coupler with oxidised colour developer is outlined below, namely

$$\text{Coupler} + \text{oxidised colour developer} \rightarrow \text{dye}$$

The coupler reaction site may contain a hydrogen atom or a functional group that is displaced during the above reaction. If there are no functional groups at the

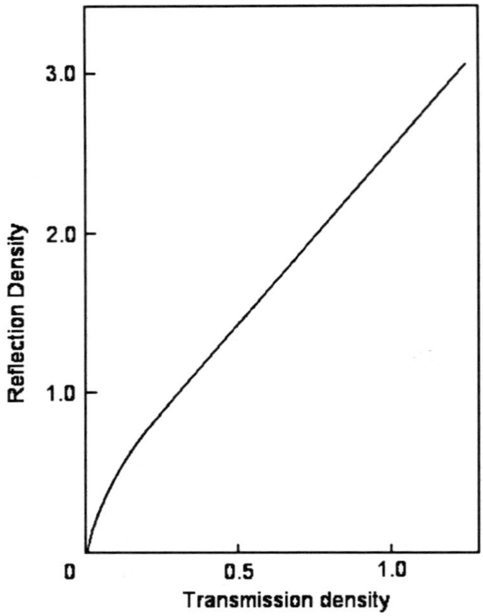

Figure 2 *A typical D_R/D_T curve*

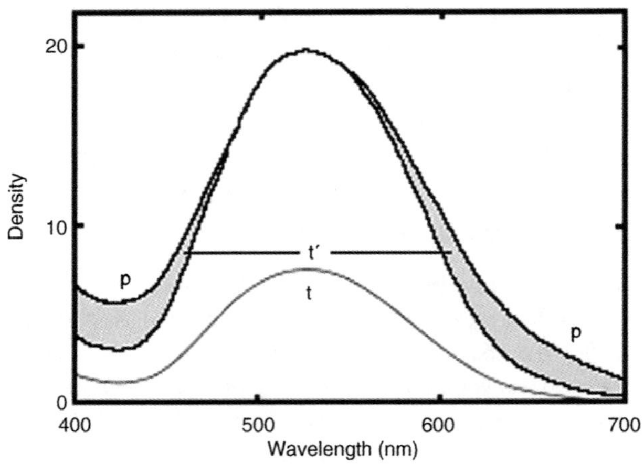

Figure 3 *Half band widths of transmitted and reflected dyes*

coupling site then the formation of dye requires 4-equivalents of silver, Figure 4.

If, however, there is a suitable non-photographic coupling-off group, then the coupler only requires 2-equivalents of silver, Figure 5.

For the use of suitable coupling-off groups, see Figure 6.

In the main, the economics of film design precludes the use of 4-equivalent couplers as the film or paper would require a higher silver

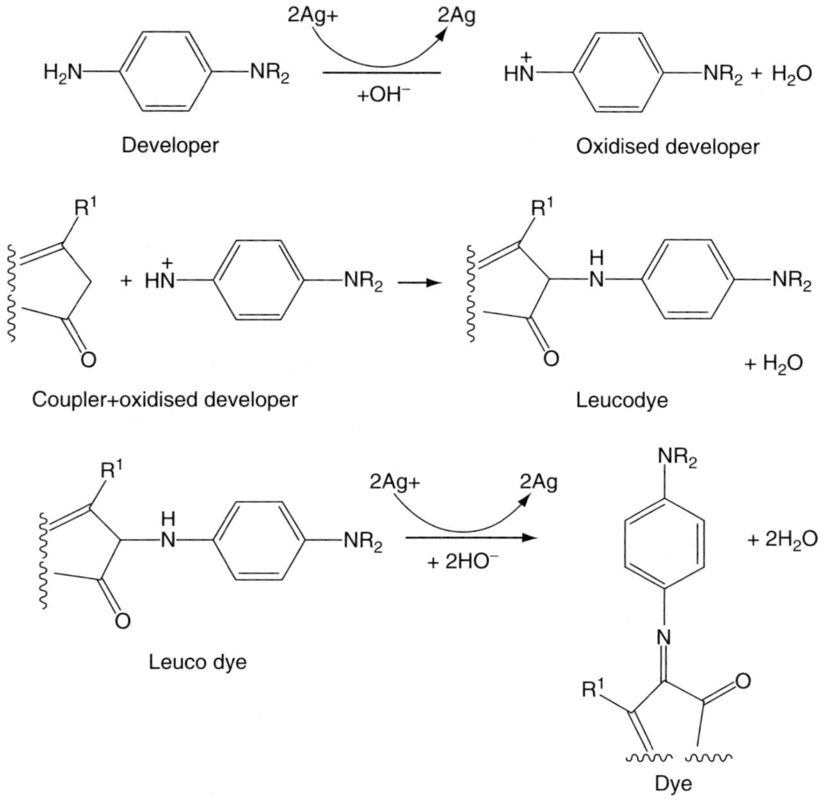

Figure 4 *The role of silver in the development process*

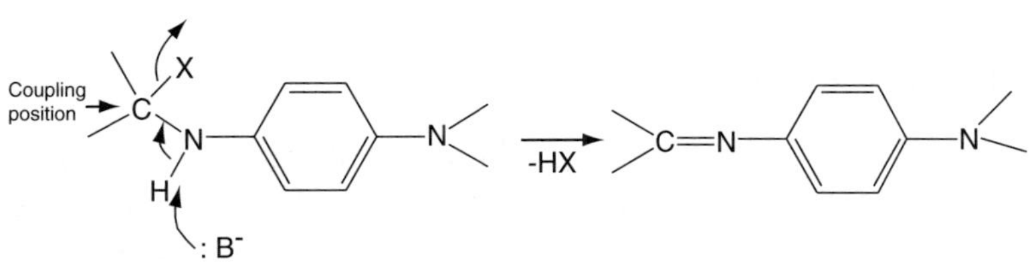

Figure 5 *Dye formation from a leuco dye*

laydown than would a corresponding film or paper with 2-equivalent couplers. The design of couplers with different coupling-off groups therefore adds to the challenge of coupler design. In some cases, the selection of the coupling-off group is based upon the reaction kinetics with oxidised colour developer, cost of intermediates, *etc.* In some specific cases, the coupling-off group is chosen so that it can be used in the film design to affect the silver development in adjacent layers. These couplers are known as either development inhibitor releasing

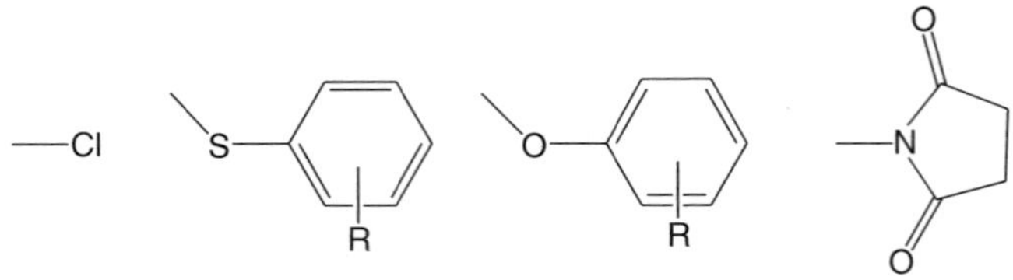

Figure 6 *Examples of coupling-off groups*

(DIR) couplers or development inhibitor anchimerically releasing (DI(A)R) couplers. These types of compounds are discussed in Chapters 8 and 9.

Additional important criteria in coupler design that affect the curve shape and λ_{max} of the resultant dye are

- the developer
- the coupler solvent used in the dispersion
- the ballast moiety.

The coupler chemist will be aware of all of these factors and will work to an aim curve for the resultant dye that is relevant for the eventual application of the coupler. During the investigations for novel couplers, many different compounds are often prepared with different side chains and functional groups.

6.2 Cyan Couplers

For the reasons mentioned above, there are many potential aim curve shapes that might be needed for a photographic product. Figure 7 details a typical film cyan dye curve that is used here to represent the type of dye curve shape that might be desired.

Kilminster and Hoke[2] prepared a range of cyan forming couplers, which they reported in their 1988 patent, Figure 8.

The synthesis of one of the couplers followed the route outlined in Figure 9.

Where 'bal' is a typical ballast, some formulae for which will be described later in this chapter.

Phenols have been the typical cyan couplers of choice for the Eastman Kodak Co. films and papers for many years. More recently, some of the scientists at Eastman Kodak Co. have made a variety of compounds in the same vein which have been combined with a stabiliser, see for example

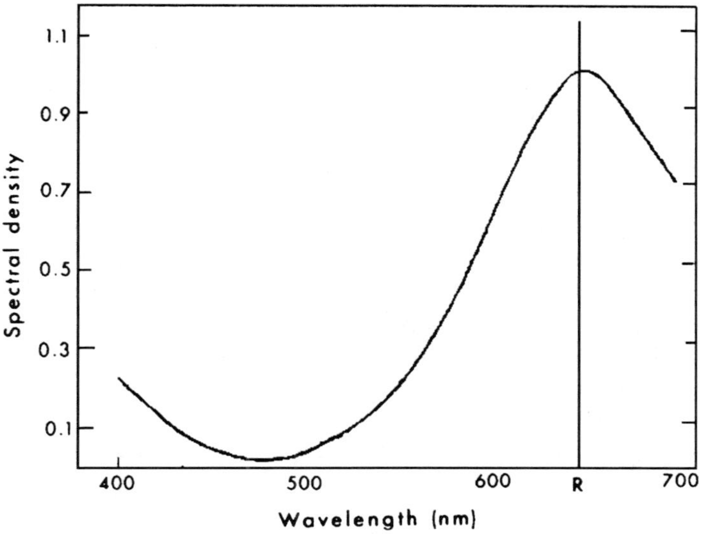

Figure 7 *A typical cyan dye curve used in film products*

the later work of Clarke *et al.*[3] from 2002. Two representative examples of the types of cyan couplers described by Clarke *et al.*, Figure 10.

These couplers are coated with a stabiliser an example of which is given below as Figure 11.

The above couplers and stabiliser were co-dissolved in high boiling coupler solvents of the type shown in Figures 12, 13 and 14.

More details concerning coupler solvents are provided in Chapter 8 during the discussion of the preparation and use of coupler dispersions. Clarke *et al.* claim that this combination of coupler, stabiliser and coupler solvent offer

> *. . . a cyan dye-forming formulation which can provide an image dye of good hue and further improved light stability without significant degradation of other photographic properties . . .*

While the Eastman Kodak Co. scientists have concentrated on the use of phenols as cyan couplers over the past few years, the most prevalent coupler type for most of the photographic manufacturers (including Eastman Kodak Co.) in earlier years involved the use of naphthols. Couplers based on the naphthol system continue to be the subjects of patents in the modern literature, although the couplers described in these patents are usually for development inhibitor applications – see Chapter 9 for a description of development inhibitors.

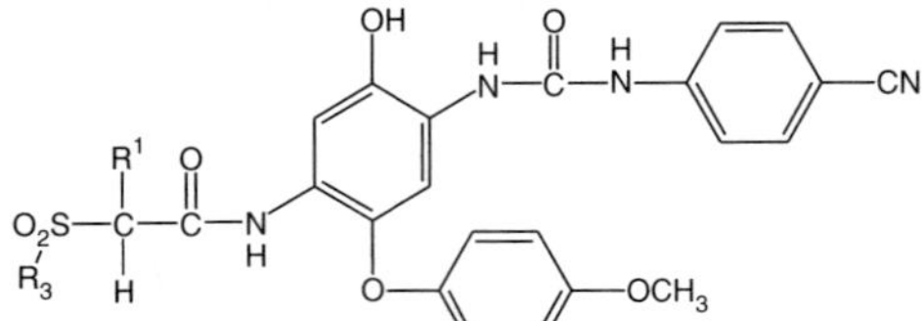

R^1	R^3	λ_{max} (nm)	HBW(nm)
$-CH_3$	$-C_{16}H_{33}$	615	145
$-C_2H_5$	$-C_{16}H_{33}$	694	118
$-C_3H_7^i$	$-C_{16}H_{33}$	697	130
$-C_{14}H_{29}$	$-CH_3$	691	128
$-C_4H_9$	$-C_{18}H_{37}$	684	122
$-C_2H_5$	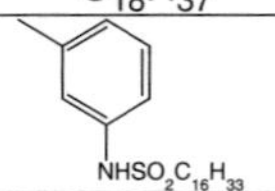	695	122
$-C_{10}H_{21}$	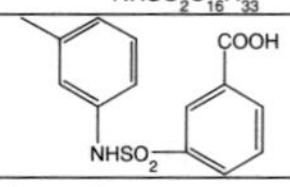	690	128
$-C_{10}H_{21}$	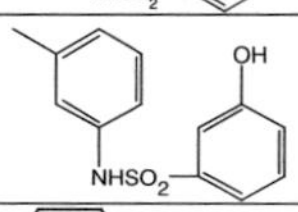	692	134
$-C_2H_5$	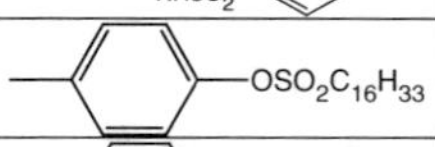	687	108
$-C_2H_5$	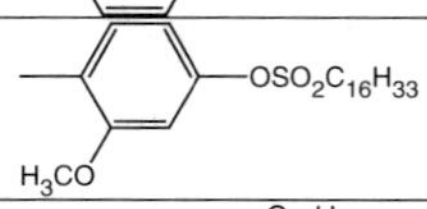	681	133
$-C_2H_5$	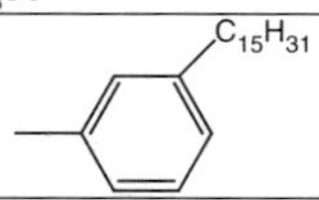	695	96

Figure 8 *Examples of cyan dye forming couplers*

Figure 15 shows a typical synthetic route to this class of cyan couplers.[4]

In their 1995 patent,[5] Takada and Okazaki of the Fuji Photo Film Co. Ltd., commented

> *...it is known that conventional phenol series or naphthol series couplers have problems in forming a cyan dye image with a usual light-sensitive material for a colour print, since they have side-absorptions in the green and blue regions. Therefore, they are not preferred in reproducing the said colours, and a solution thereof has been desired. An example of an unnecessary side-absorption deteriorating the colour reproduction is that the green colour of, for example, a green leaf which is photographed on a colour negative film and reproduced on a colour print inclines to a brownish green colour in some cases. As the means for solving this problem, 2,4-diphenylimidazoles described in*

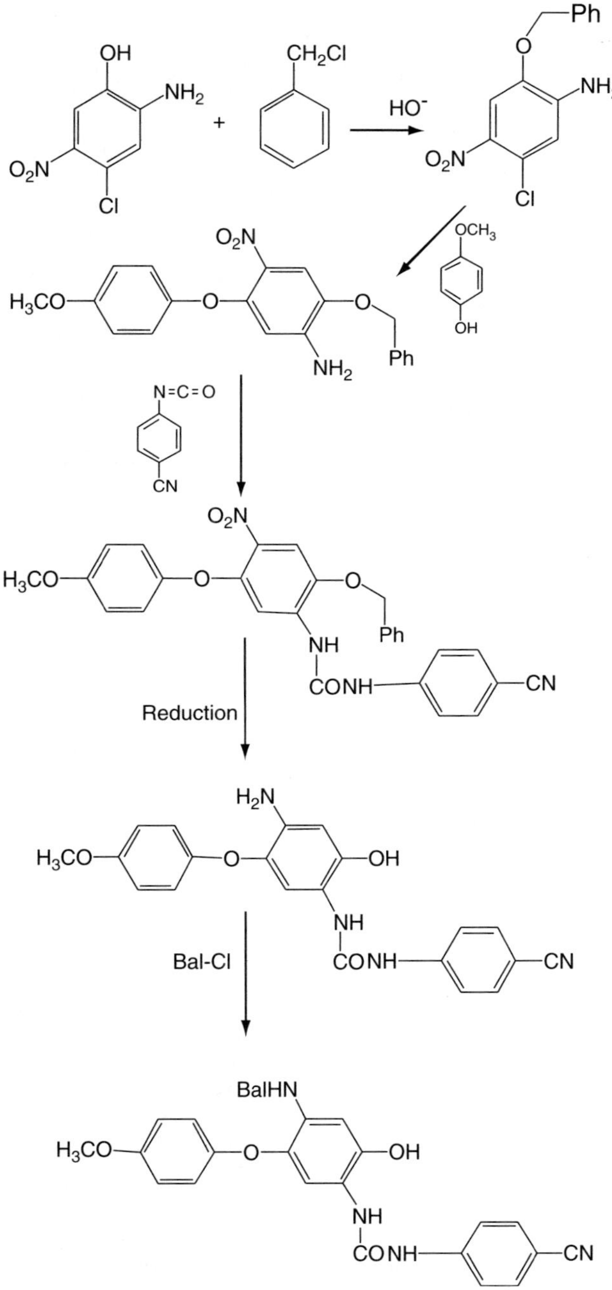

Figure 9 *A route to a cyan coupler*

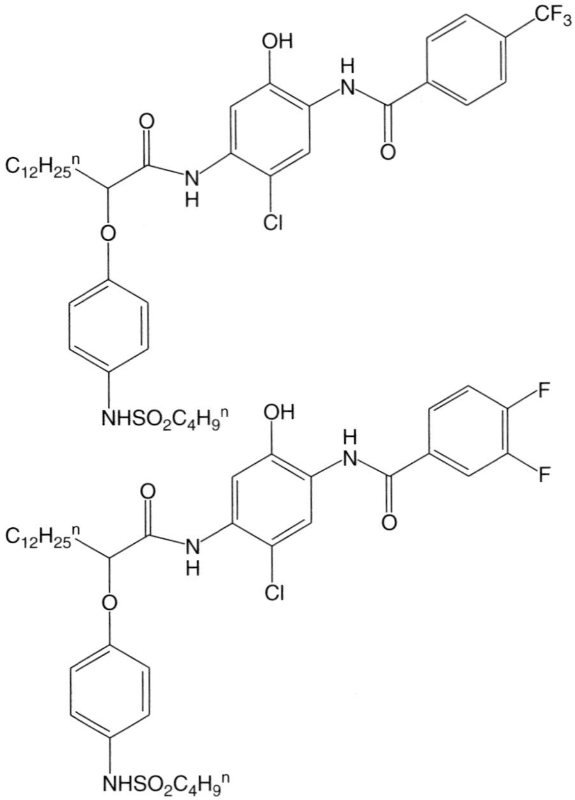

Figure 10 *Typical cyan couplers*

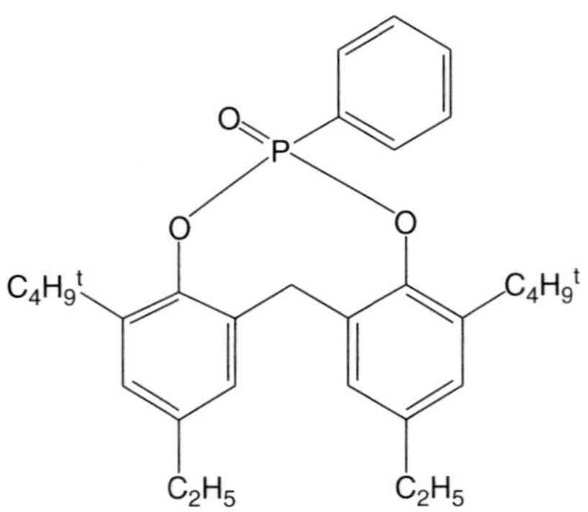

Figure 11 *An example of a stabiliser for use with cyan couplers*

European Patent 294,453A2 are proposed. In the dyes formed by these couplers, undesirable side-absorption in short wavelength regions is reduced compared with that of conventional dyes. Accordingly, they are preferable in terms of colour reproduction.

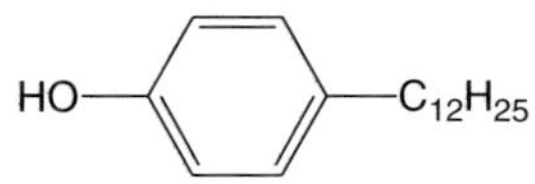

Figure 12 *A high boiling coupler solvent*

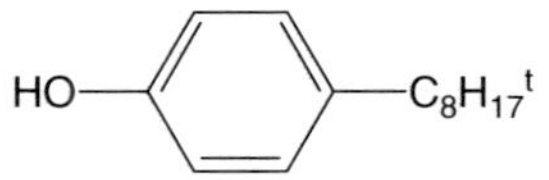

Figure 13 *A typical coupler solvent*

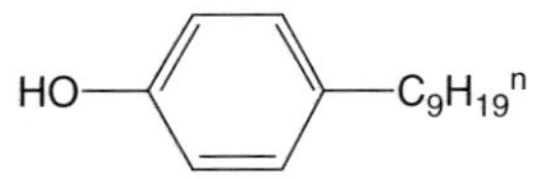

Figure 14 *A commercial coupler solvent*

> *However, in these couplers, the colour reproduction is not necessarily sufficient and additionally, there remain the problems that in practical use coupling activity is low and that fastness to heat and light is poor. Accordingly, they are impractical. The pyrazoloazole series couplers described in JP-A-64-553 (the term "JP-A" as used herein means an unexamined published Japanese patent) are improved as to side-absorption in short wavelength regions compared with that of conventional dyes, but those couplers do not always have sufficient colour reproducibility and there remains the problem that colour developing performance is notably low ...*

Figure 16 details the general types of heterocycles that are described in the patent, and Figure 17 details a specific synthesis.

6.3 Magenta Couplers

A typical film dye curve for a magenta dye is presented as Figure 18. Dyes that approximate this curve shape have been generated from several different nitrogen containing heterocyclic nuclei by all of the photographic manufacturers. Commercial couplers are usually derivatives of the pyrazolone or pyrazolotriazole ring systems. Some other ring systems that have been investigated as potential magenta couplers are provided for completeness.

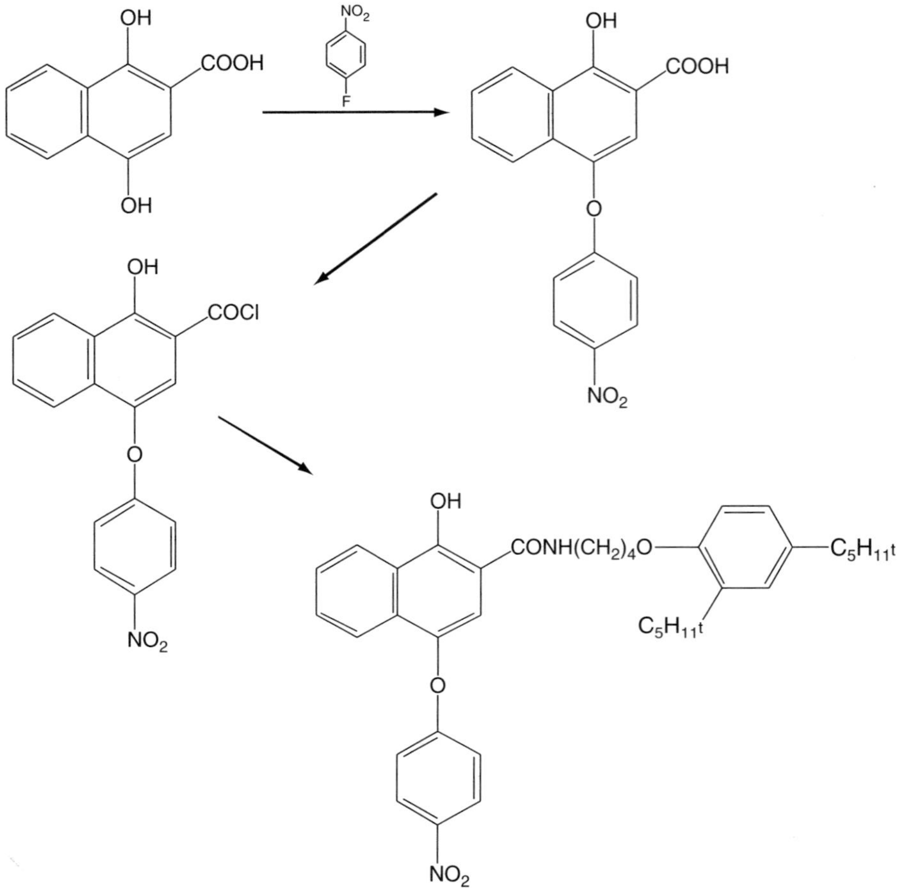

Figure 15 *A route to cyan couplers based on naphthols*

6.3.1 The Pyrazolone Nucleus

In their 1951 paper, Vittum *et al.*,[6] described the effect on dye hue of electron donating or withdrawing groups, Figure 19.

When Y is an electron-donating group, the dye λ_{max} is shifted to shorter wavelength. This is known as a hypsochromic shift. When Y is an electron-withdrawing group there is a corresponding shift to deeper wavelengths, known as a bathochromic shift.

The data in Figure 19 is typical of studies undertaken on a new ring system. The only mechanism open to the chemist designing new couplers is to manipulate the position or type of the substituents in the coupler molecule as the developer is fixed for a given application. In general pyrazolone dyes suffer from an unwanted secondary absorption in the 440–460 nm range, which if left uncorrected, would affect the colour balance

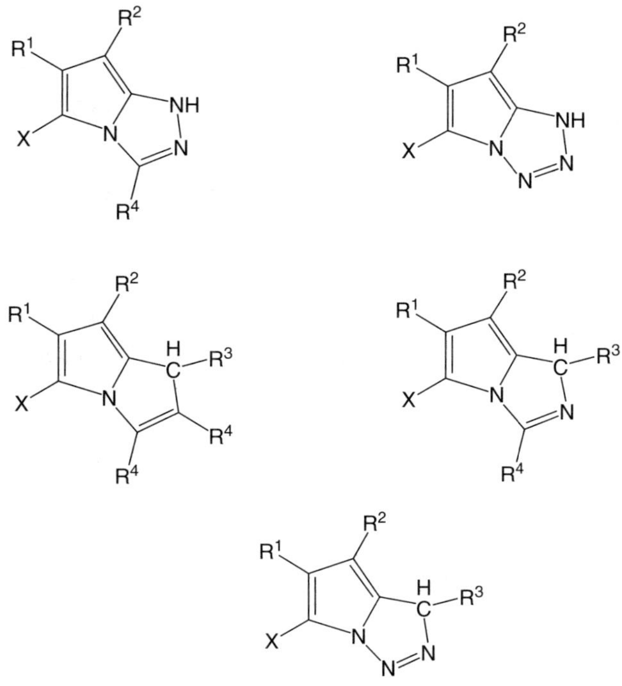

Figure 16 *Cyan dye forming pyrazoloazole couplers*

of the final film or paper. When Y is a nitrogen-donating group, for example an amido or amino group, considerably reduced unwanted blue absorptions are observed. This has led to large studies of electron donating, nitrogen containing-groups as the Y substituent, see for example.[7–10]

Many routes have been developed for the synthesis of pyrazolone couplers. Weissberger[11] provided an early route, Figure 20.

While the majority of the early commercial magenta couplers were based on the pyrazolone ring system, latterly the most prevalent magenta couplers have been based on the 1H-pyrazolo[3,2-c]-s-triazole system. This ring system has been described above for use as a cyan coupler. The use of the ring system for magenta couplers is more widespread. Figure 21 shows a synthesis from the Browne and Normandin[12] patent, which is reproduced here to reflect a more modern synthesis of this magenta coupler type.

Earlier patents concerning the 1H-pyrazolo[3,2-c]-s-triazole system, see for example ref 13–16, describe other synthetic pathways. Many examples of this coupler class have been prepared by all of the photographic manufacturers.

The imidazo[1,2-b]pyrazole nucleus was first described by Rogers and Bailey[17] of the Eastman Kodak Co. and was prepared by the following scheme (Figure 22) some examples of which are shown in Figure 23.

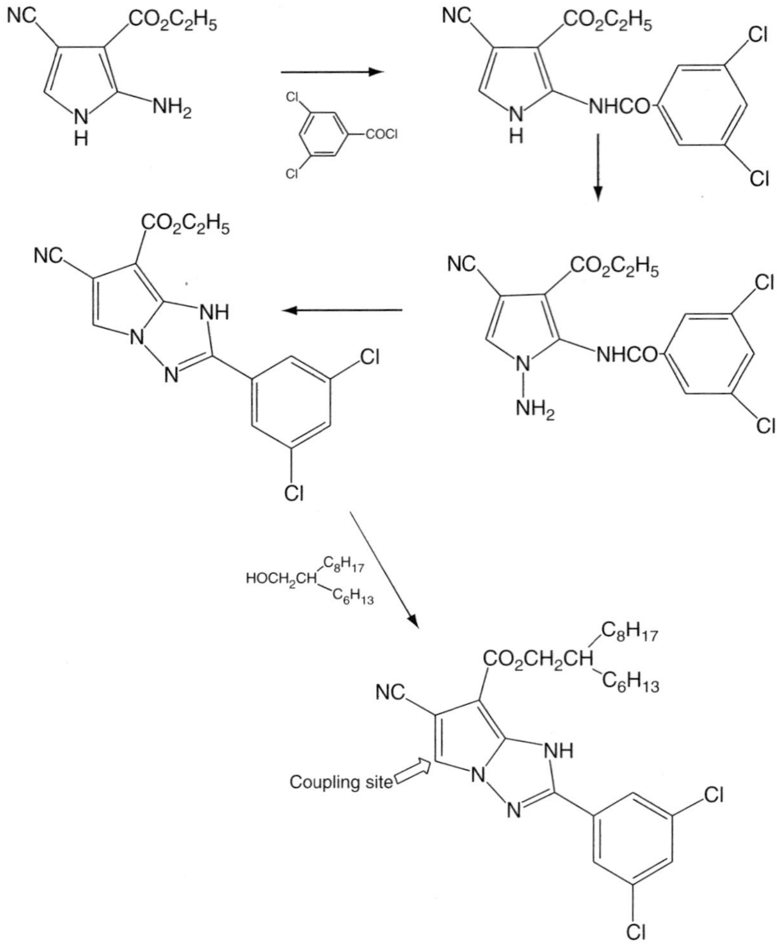

Figure 17 *The synthesis of a pyrazoloazole coupler*

The indazole ring system has been investigated by Ilford Ltd.[18] Figure 24 provides the synthesis.

The resultant dye which is formed with oxidised colour developer is zwitterionic and requires four silver equivalents, Figure 25. In most cases indazolone couplers are prone to form an undesirable yellow dye in the bleach step of the processing solutions.

6.4 Yellow Couplers

Figure 26 details a yellow dye curve shape, representative of the curve type needed for a photographic product. In this case, the curve is more suitable for a colour film application.

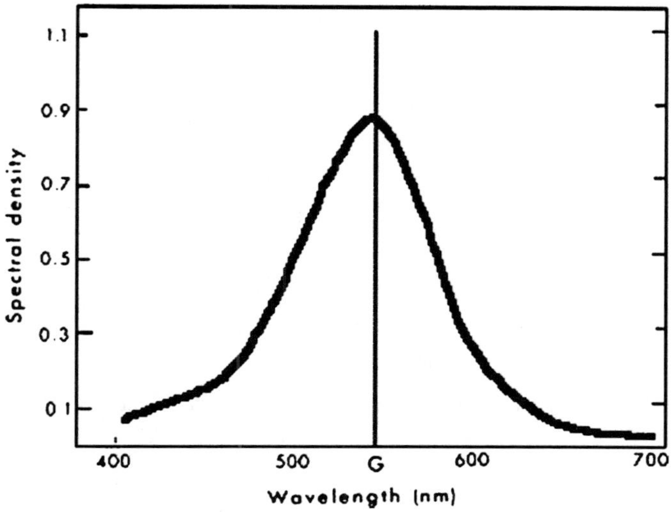

Figure 18 *A typical magenta dye curve shape*

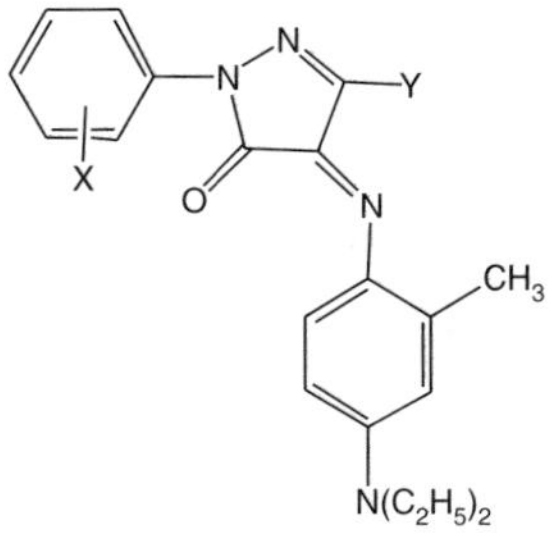

Magenta Azomethine Dyes

X	Y	λ_{max} nm[a]	E_{max}x 10^{-4} [a]
H	H	546	3.3
H	CH_3	522	3.6
H	C_6H_5	537	3.1
H	NH_2	506	4.1
H	NHC_6H_5	522	4.7
H	$NHCOC(C_2H_5)HC_6H_5$	526	5.2
H	$CONH_2$	565	3.8
4-NH_2	CH_3	514	3.6
4-Cl	CH_3	526	3.9
2,4,6-tri-Cl	CH_3	530	4.1

[a] measured in butyl acetate solution

Figure 19 *Substituent effects on λ_{max} and extinction coefficient*

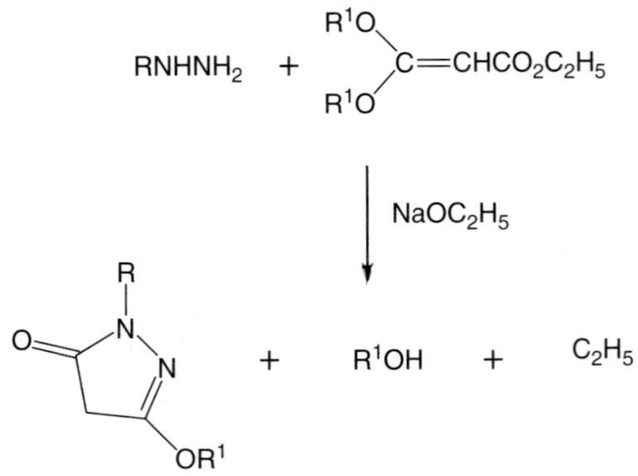

Figure 20 *An early synthesis of a pyrazolone 4-equivalent coupler*

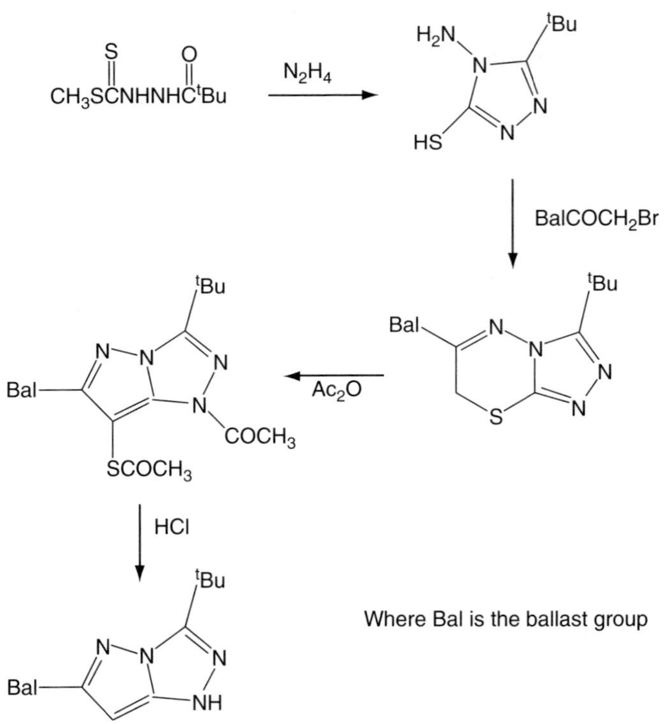

Figure 21 *A relatively recent pyrazolo[3,2-c]-s-triazole synthesis*

The most important classes of yellow couplers are derived from β-ketocarboxamides, Figure 27.

The classic form of these couplers, used in all products worldwide until the late 1960s to the early 1970s, was the benzoylacetanilides, Figure 28.

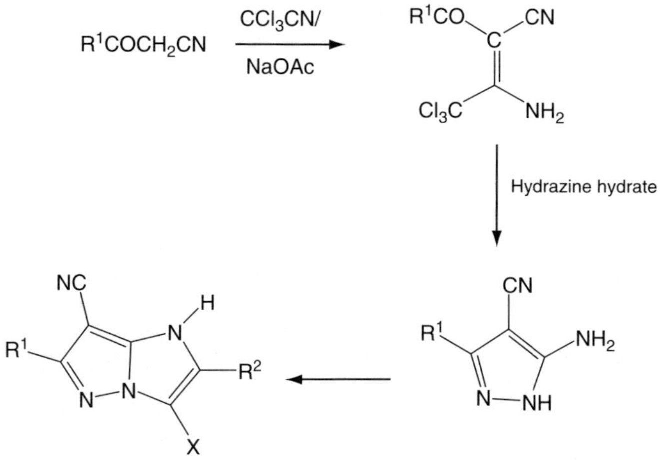

Figure 22 *A route to imidazo[1,2-b]pyrazoles*

Some of the seminal work on these coupler types was reported by Weissberger *et al.*[19] in 1957, where they provided a table of structure/dye hue data. Figure 29 shows that changes in X to an electron donating group causes a hypsochromic shift, whereas electron withdrawing groups cause bathochromic shifts.

Many patents were filed concerning novel benzolyacetanilide couplers, including Hunt of Ilford[20] who claimed some novel ballast groups; see later section in this chapter. A typical synthesis of a benzolyacetanilide coupler is shown in Figure 30. More details concerning this coupler and some analogues appear in.[21]

Although of commercial importance, benzoylacetanilide couplers suffered some drawbacks. Most of the research attention was diverted to pivaloylacetanilides in the last 1960s, Figure 31, as they were shown to be more favourable than benzoylacetanilide dyes for the following criteria

- sharper-cutting dyes
- less unwanted green absorption
- improved light stability.

The 4-equivalent parent pivaloylacetanilide coupler shown in Figure 31 is less active towards oxidised colour developer than is the equivalent benzoylacetanilide coupler. Fortunately, the 2-equivalent analogue pivaloylacetanilide couplers are of sufficient activity for them to be couplers of choice for the last 30 years.

Figure 32 shows the synthesis of a 4-equivalent pivaloylacetanilide coupler. This synthetic route is extremely versatile and can be used for a

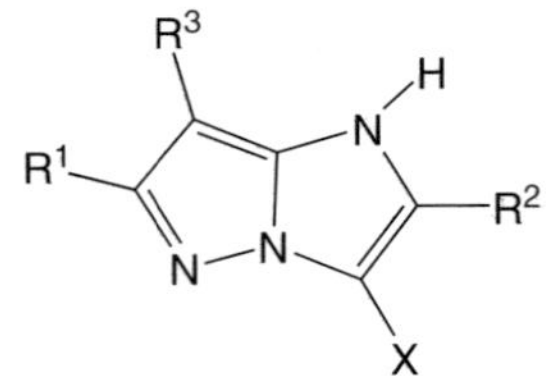

Coupler No.	X	R^1	R^2	R^3
1	-H	$-^tBu$	-Me	-CN
2	-H	$-C_6H_5NO_2$	$-^tBu$	$-CO_2Et$
3	-H	-H	$-^tBu$	-CN
4	-H	-H	$-^tBu$	$-CONH_2$
5	-H	-H	$-^tBu$	$-CO_2Et$
6	-OPh	$-C_6H_5NO_2$	$-^tBu$	$-CO_2Et$
7	-H	$-C_6H_5NH_2$	$-^tBu$	-CO-Et

Figure 23 *Several examples of imidazo[1,2-b]pyrazole couplers*

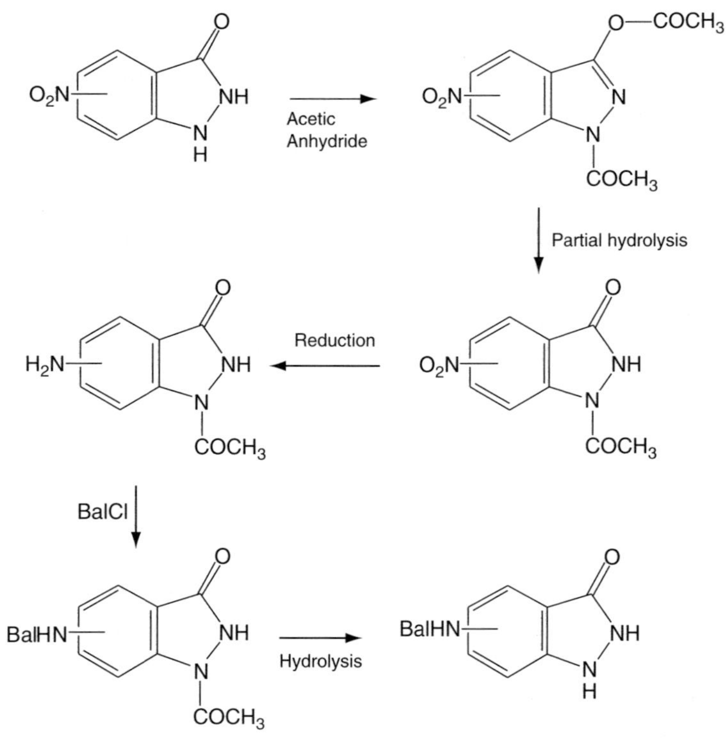

Figure 24 *A synthesis of the indazole ring system*

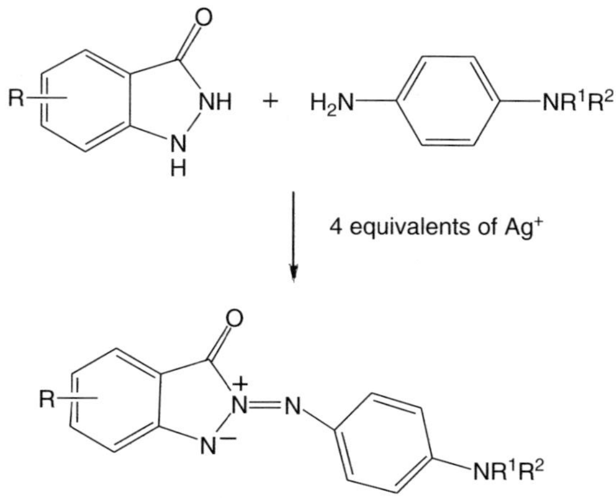

Figure 25 *A zwitterionic dye derived from an indazolone*

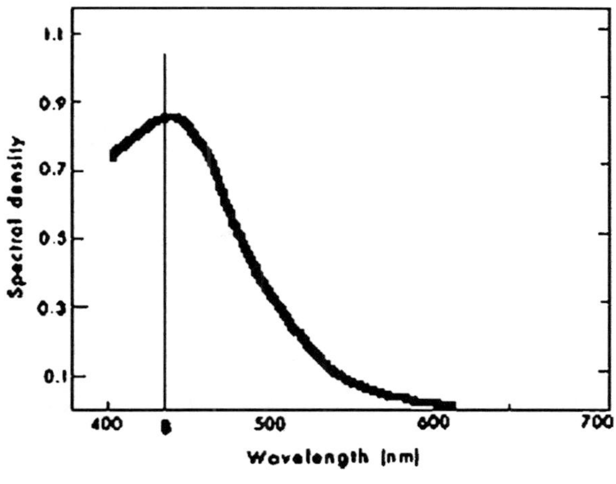

Figure 26 *A representative curve shape of a yellow dye*

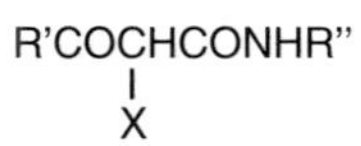

Figure 27 *The general structure of a β-ketocarboxamide*

wide variety of 2- and 4- equivalent yellow couplers with a variety of different ballast groups.[22]

6.4.1 Ballast Groups

Some mention has been made of the term a 'ballast group' or 'ballast'. As the name implies these moieties are designed to act as an anchor or

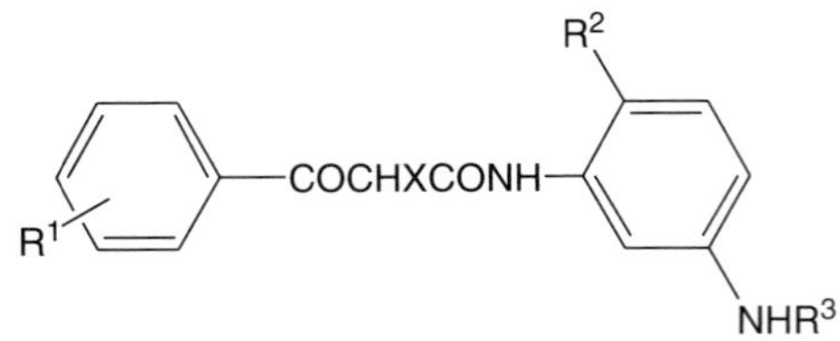

Figure 28 *A 2-equivalent benzolylacetanilide coupler*

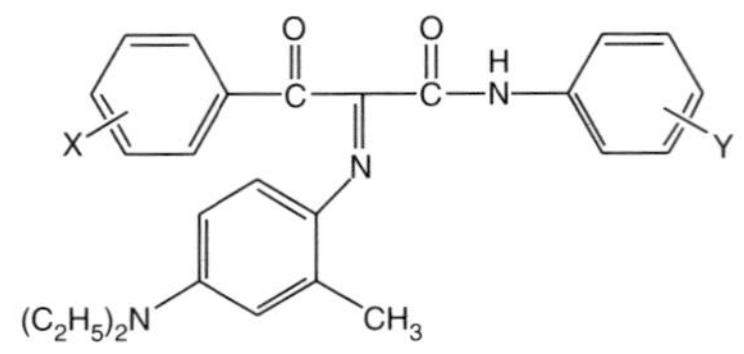

X	Y	λ_{max} nm[a]	E_{max} x 10^{-4} [a]
H	H	433	1.6
o-OCH_3	H	420	1.0
p-OCH_3	H	430	1.6
p-Cl	H	436	1.6
H	o-OCH_3	432	1.9
H	p-OCH_3	430	1.5
H	o-Cl	442	2.2
H	p-Cl	438	1.7
H	o-NO_2	460	1.9
H	p-NO_2	451	2.1
H	o-$NHCOC_6H_5$	438	1.7
H	p-$NHCOC_6H_5$	435	1.6

[a] measured in butyl acetate solution

Figure 29 *Substituent effects on dye hue for benzolylacetanilide dyes*

ballast for the coupler during the specific stages of both manufacture of the film or paper and during the processing stage. Figure 33 shows the potential effect of an un-ballasted coupler molecule and one that has a ballast.

Figure 33 represents the effect of water, perhaps from the processing solutions. In the diagram, the coupler molecules that have no ballast moiety are washed nearer to the air/gel surface by the water, whereas the ballasted couplers remain in position. The extent of the shift has been exaggerated to demonstrate that the dye molecules, which provide the final image, will be easily displaced without a ballast group. Several consequences are likely. The worst case would be that a dye formed from one coupler is washed into an adjacent layer of a different colour record, which would cause many potential problems of colour reproduction and sharpness.

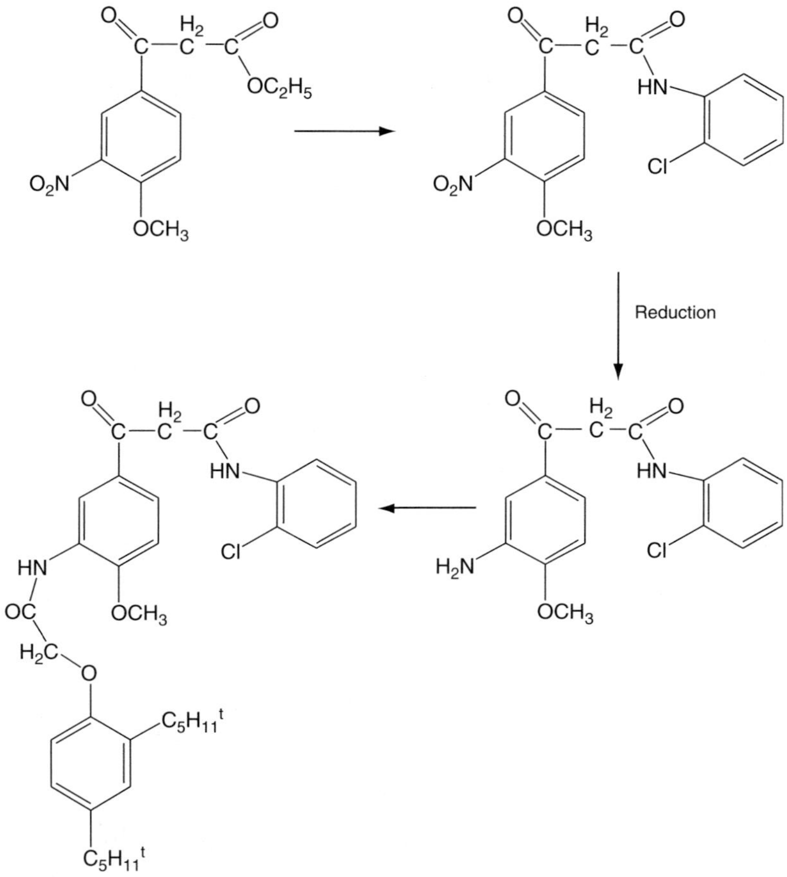

Figure 30 *A synthetic route to 4-equivalent benzolylacetanilide couplers*

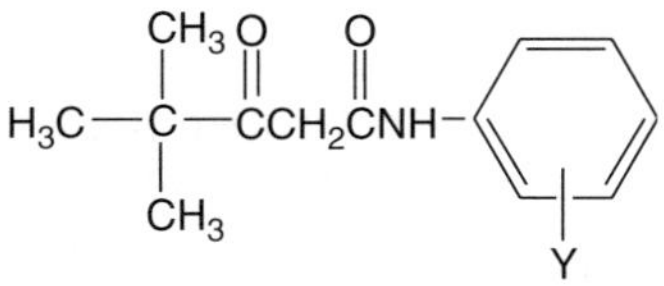

Figure 31 *A 4-equivalent pivaloylacetanilide coupler*

All the photographic manufacturers use ballasted couplers or polymeric couplers (see the next section). Figure 34 shows some typical ballast moieties.

These examples appear in patents from most of the photographic manufacturers. It is not an exhaustive list. Ballasts can affect the

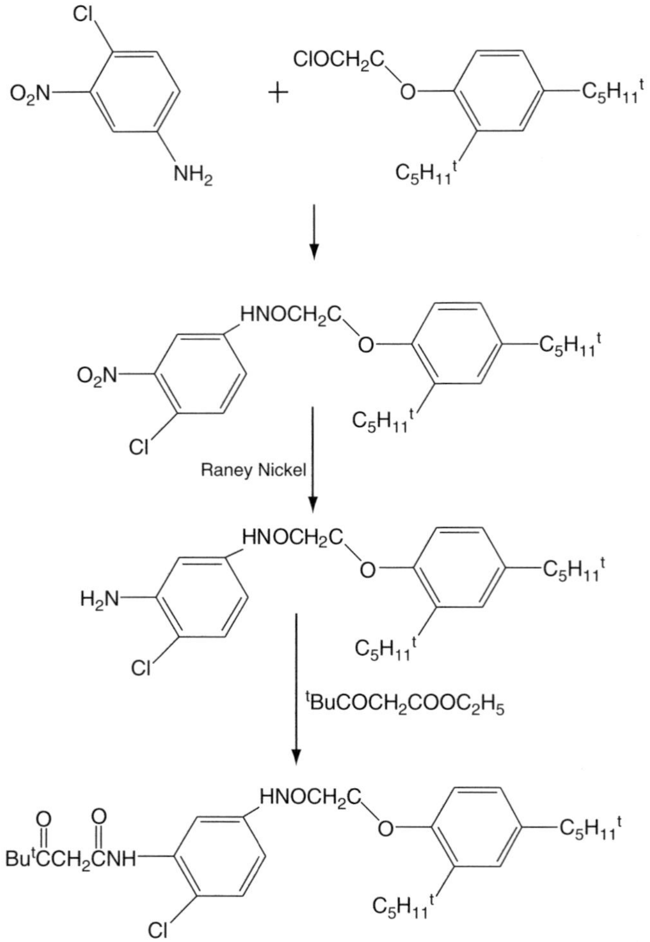

Figure 32 *The synthesis of a 4-equivalent pivaloylacetanilide coupler*

planarity of the dye moiety of the molecule, which can in turn affect both the λ_{max} of the dye and the effect of oxygen or light on the dye which is manifested in the stability of the dye. For these and other reasons, ballasts play an important part in the final coupler synthesis.

6.4.2 Polymeric Couplers

Arguably, the natural extension of coupler moiety attached to ballast might be to attach two, three or more, in other words the synthesis of polymeric couplers. Although polymeric couplers have been used in commercial products, their introduction only took place in the 1980s. These polymeric couplers were exclusively magenta dye forming

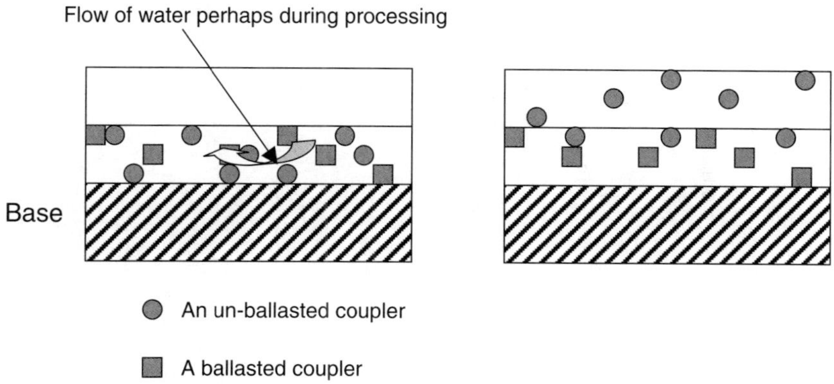

Figure 33 *The effect of a ballast moiety on the movement of a coupler molecule*

couplers. The issue of most concern was the increase in sharpness of the underlying layers, and not the synthesis of the various ballast moieties. Consider Figure 35.

The left hand diagram in Figure 35 represents the layers of magenta dye forming couplers containing conventional ballasted couplers coated in a film format. The upper layers have been omitted from the diagram for simplicity. The right hand diagram in Figure 35 represents the thinner layers one might expect from coating polymeric couplers. Red light would be scattered by the silver halide crystals in the magenta forming layer but not be absorbed. It might be the case that a silver halide crystal in the cyan dye-forming layer reflects the light which is then subsequently captured by a silver halide crystal in the same layer. This reflection might also occur in the lower of the two cyan dye-forming layers. In either event the back reflected light has the potential to travel a greater distance from the desired exposure point in the left hand diagram compared with the right hand diagram. This is because the magenta dye forming layers are thinner in the right hand case, which alters the angle of incidence of the light entering the cyan layers.

The ability to control – or at least influence – the extent of the red sharpness by making changes to the magenta chemistry provides the film builder with a degree of flexibility previously unavailable. It was for this primary reason that polymeric couplers were introduced. One might consider that yellow forming polymeric couplers would be of greater effect. In terms of the distance between the top of the film layers and the cyan forming layers that is a correct statement. However, the emulsions used in the yellow dye forming layers tend to be larger t-grain type emulsions, which affect the optical path of the red light in ways that are

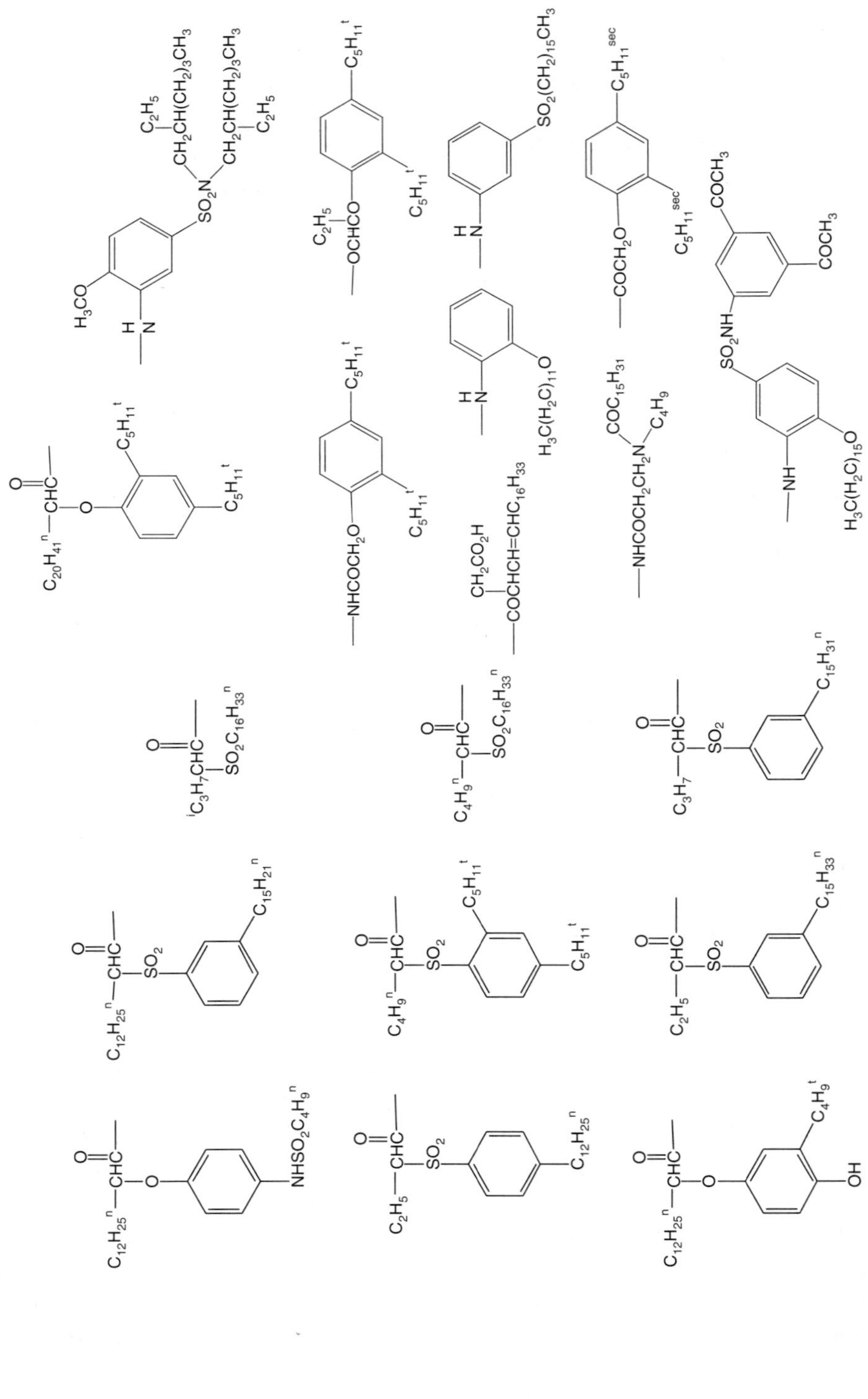

Figure 34 *A selection of ballast groups from different manufacturers*

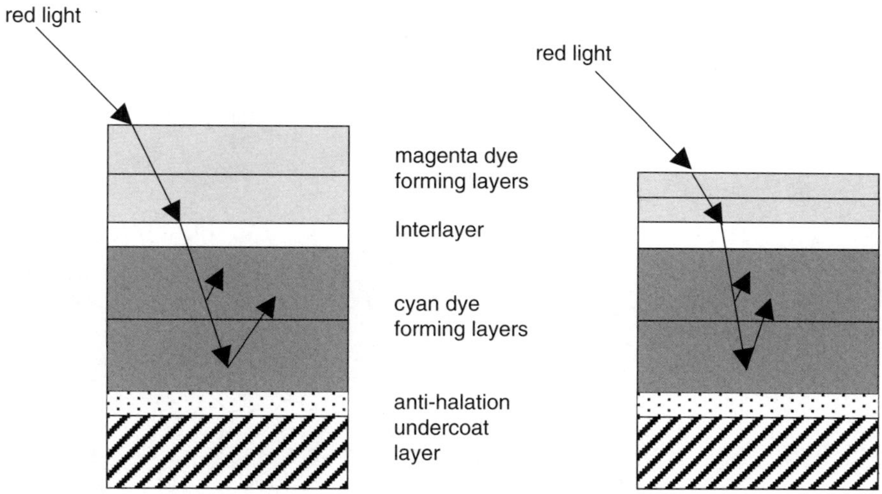

Figure 35 *The effect of thinning upper layers on light reflection*

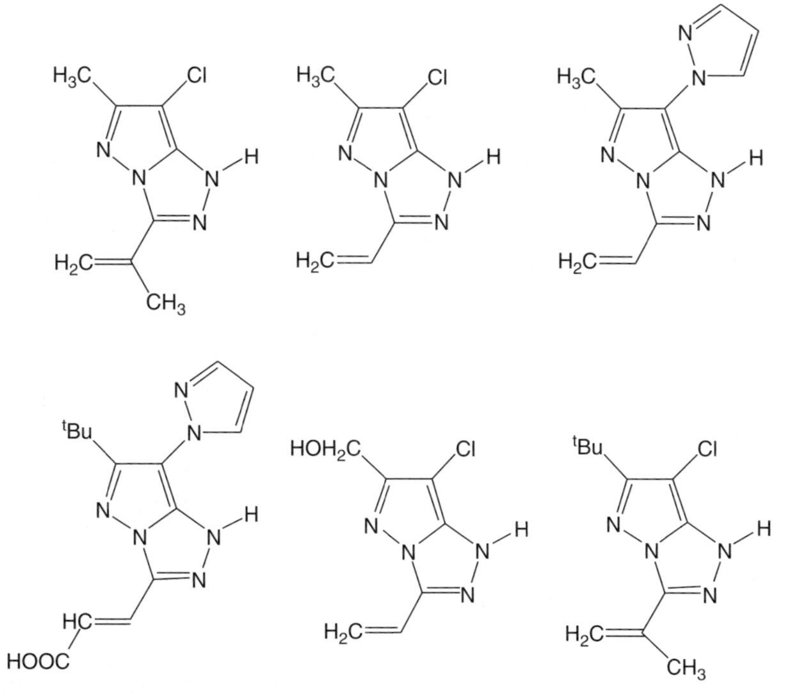

Figure 36 *Magenta coupler monomers*

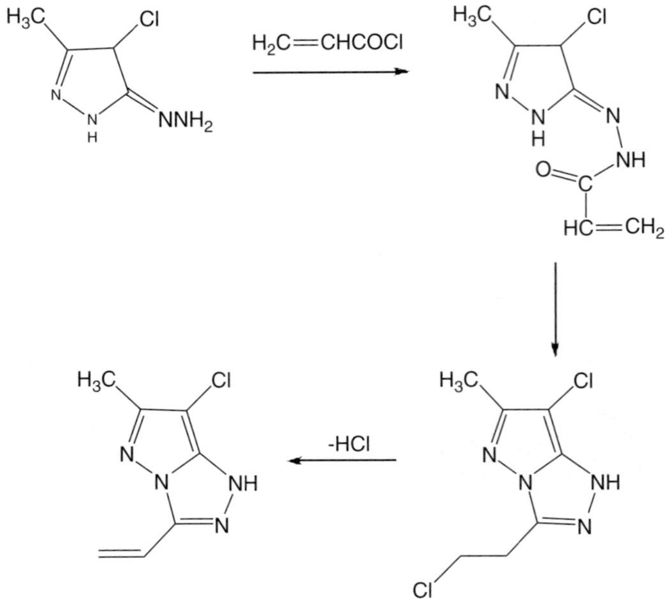

Figure 37 *The synthesis of a magenta coupler monomer*

different to the thicker grained emulsions that are often used in the magenta dye forming layers. The need for polymeric yellow dye forming couplers is therefore not as great as for magenta dye forming polymeric couplers.

Arguably, Agfa-Gevaert AG have published more patents than their competitors, concerning polymeric couplers – see ref 23. Helling, Biialek and Weimann published structures of some of the coupler monomers, Figure 36. They polymerised these and other monomers with a combination of *n*-butyl acrylate and divinylbenzene, styrene and methyl methacrylate, or methyl acrylate and butyl methacrylate, for example.

Figure 37 outlines the synthesis of a typical magenta dye forming monomer.

Eastman Kodak Co. favour the use of the following coupler monomers, Figure 38.

The magenta dye forming monomers are co-polymerised with non-photographically useful monomers some of which are presented in Figure 39.

These materials can be polymerised using the method described by Tien-Teh Chen, S.W. Cowan and E. Schofield.[24]

> *... the mixture of monomers which is polymerized to form the polymeric couplers of the present invention includes the ionic monomer in an amount*

sufficient to provide the polymeric coupler with less than 10 weight percent of the ionic monomer. Preferably, the ionic monomer is included in an amount sufficient to provide the polymeric coupler with from 1 to 5 weight percent of the ionic monomer. In a further preferred embodiment, the polymeric coupler contains from 10 to 90 percent by weight of the coupler monomer, more preferably from 30 to 70 percent by weight of the coupler monomer, not greater than 10 percent by weight of the ionic monomer and a remainder of one or more non-dye-forming ethylenically unsaturated monomers. The average particle size of the water-dispersed polymeric couplers according to the present invention is generally in a range of from about 20 to 200 nanometers, depending on the amount of ionic monomer included therein. Additionally, the polymeric couplers preferably have a weight average molcular weight of from 2000 to about 40,000 . . .

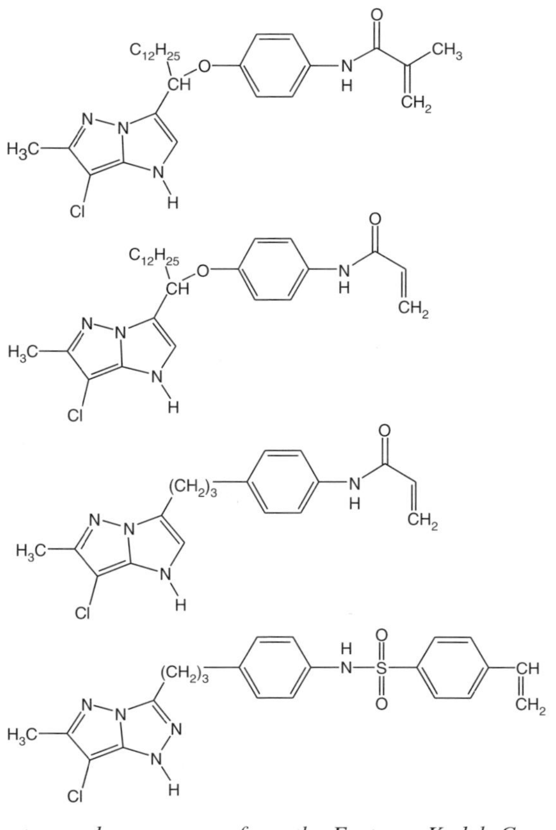

Figure 38 *Magenta coupler monomers from the Eastman Kodak Co*

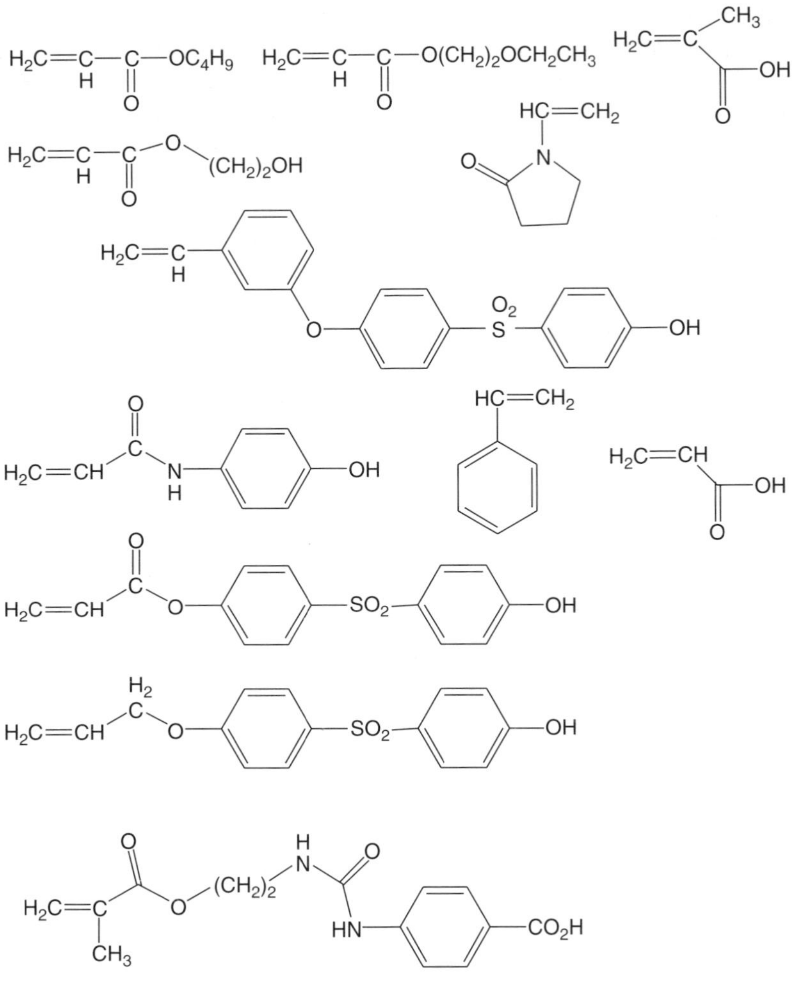

Figure 39 *Non-photographically active monomers*

References

1. R.W.G. Hunt, *The Reproduction of Colour*, 4th edn, Fountain Press, Kingston-upon-Thames, 1987, ISBN 0-85242-356X, p. 225.
2. K. Kilminster and D. Hoke, US 4,775,616, Eastman Kodak Co.
3. D. Clarke, L.J. Leyshon and K.E. Smith, EP 1,197,798, Eastman Kodak Co.
4. GB 1,084,480, Eastman Kodak Co.
5. K. Takada and K. Okazaki, US 5,397,691, Fuji Photo Film Co. Ltd.
6. G. Brown, B. Graham, P. Vittum and A. Weissberger, *J. Amer. Chem. Soc.*, 1951, **73**, 919.

7. H. Porter and A. Weissberger, US 2,343,703, Eastman Kodak Co.
8. S. Popeck and H. Schulze, US 2,829,975, General Aniline and Film Corp.
9. I. Salminen, C. Barr and A.Loria, US 2,895,826, Eastman Kodak Co.
10. B. Graham and A. Weissberger, US 2,691,659, Eastman Kodak Co.
11. H.D. Porter and A. Weissberger, US 2,439,098, Eastman Kodak Co.
12. A.T. Browne and S.E. Normandin, US 4,992,361, Eastman Kodak Co.
13. S. Yoshimoto, S. Nakagawa, Y. Kaneko and S. Sugita, US 4,994,360, Konica Corporation.
14. T. Sato, O. Takahashi, H. Naruse and Y. Mizukawa, US 5,066,676, Fuji Photo Film Co. Ltd.
15. J. Bailey and D.A. Thomas, GB 1,398,979, Eastman Kodak Co.
16. H. Ohya and S. Kida, US 5,071,739, Konica Corporation.
17. D.N. Rogers and J. Bailey, US 4,728,598, Eastman Kodak Co.
18. GB 1,285,432, Ilford Ltd.
19. G. Brown, J. Figueras, R.J. Gledhill, C.J. Kibler, F.C. McCrossen, S.M. Parmerter, P. Vittum and A.A. Weissberger, *J. Amer. Chem. Soc.*, 1957, **79**, 2919.
20. F.G. Hunt, GB 1,240,907, Ilford Limited.
21. GB 1,261,156, Fuji Photo Film Co. Ltd.
22. A. Weissberger and C.J. Kibler, US 3,265,506, Eastman Kodak Co.
23. G. Helling, R. Biialek and R. Weimann, US 5,354,826, Agfa-Gevaert AG,.
24. Tien-Teh Chen, S.W. Cowan and E. Schofield, US 5,455,147, Eastman Kodak Co.

CHAPTER 7

Image Dye Formation and Stability

Although the technology for incorporating a coupler dispersed directly into an emulsion layer is known – see for example Agfacolor – it is more likely that the coupler is dispersed in a high boiling solvent (>150°C), and these coupler containing droplets dispersed in a gelatin solution. This mixture is often referred to as 'coupler dispersion' or simply a dispersion.

In a typical case the immiscible coupler solvent is heated with the coupler (usually each coupler is dispersed in one half its weight of solvent) until solubility is achieved. In some cases a lower boiling water soluble solvent, known as the auxiliary solvent, might be added to aid solubility. The coupler solution is then mixed with a gelatin/water solution at elevated temperatures and the mixture subject to high shear. Often a surfactant is added in order to help to break up the organic phase into sub-micron droplets.

In their 2002 patent, D. Clarke, L.J. Leyshon and K.E. Smith[1] offered a number of structures of coupler solvents, Figure 1.

These are just some of the many liquids that could be considered as coupler solvents. The list below, Figure 2, shows a representative selection of the different types used by the major photographic manufacturers. These solvents have been in use for many years, see for example ref 2. New solvents have been introduced more recently, some of which appear in Figure 2.

Figure 3, contains some known auxiliary coupler solvents

A more comprehensive list of coupler solvents, which is also cited in some of the modern Fuji patents, was published by Jelley and Vittum[3] in their 1940 patent.

Coupler solvents can affect the reactivity of the coupler, which is dissolved in them, additionally they often affect the final dye, once the coupler has reacted with oxidised colour developer. The half-band width

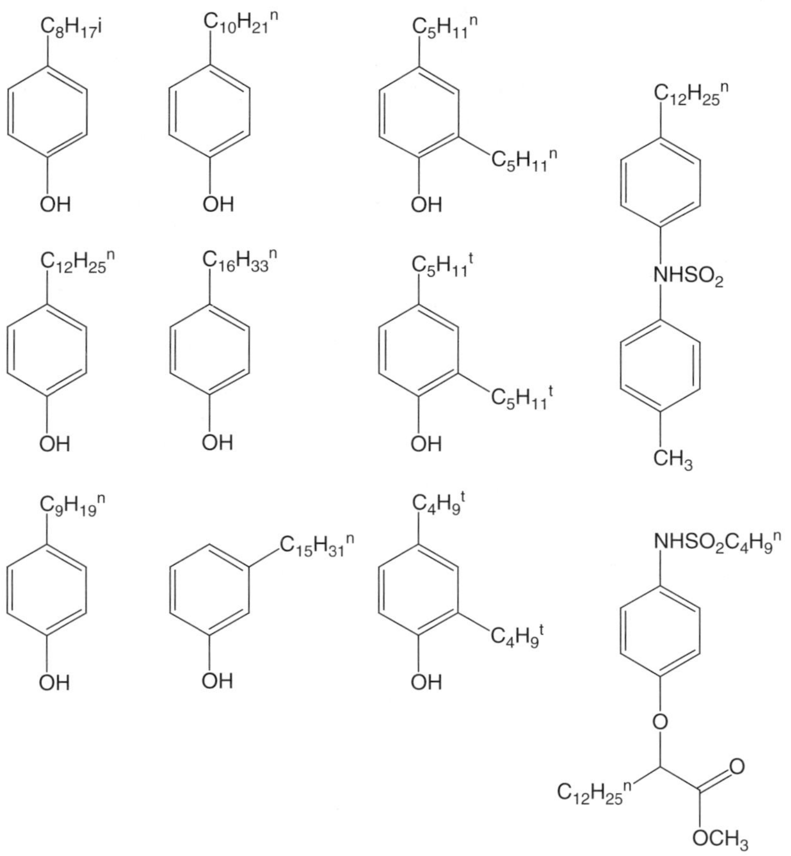

Figure 1 *Examples of coupler solvents used by the Eastman Kodak Co*

and the λ_{max} of the dye can also be affected. The design of the dispersion is therefore a crucial activity in the design of the final photographic product.

The relationship between coupler solvent and dye hue was published for some novel coupler solvents proposed by McCrossen and Osborn[4] of Eastman Kodak Co., Figure 4.

Dispersions have been made in a similar manner for many years. Any patent claiming a coupler invention will contain the method – from all of the photographic manufacturers. Thus the method described by Poslusny *et al.*,[5] in their 2004 patent, represents the standard dispersion-making method used by Eastman Kodak Co.

> *... typical comparative dispersions were prepared by adding an oil phase containing a 1:1:3 weight ratio of coupler:dibutylsebacate:ethyl acetate to*

Coupler solvents

acetyl tributyl citrate
benzyl salicylate
bis(2-ethylhexylsulfoxide)
dibutyl laurate
dibutyl phthalate
dibutyl-p-dodecylphenol
dibutylphthalate
di-n-butylphthalate
diphenyl 2-ethylhexyl phosphate
dodecylbenzene
N,N-dibutyldodecanamide
N,N-dibutyl-p-toluenesulfonamide
N-butylacetanalide
octyl benzoate
tricresyl phosphate
tricresyl phosphate
tricresylphosphate
trioctyl phosphate
trioctylphosphine oxide
tritolyl phosphate

Figure 2 *Coupler solvents used more widely in the industry*

butyl acetate
cyclohexanone
ethyl acetate
isopropanol
methyl ethyl ketone

Figure 3 *Examples of auxiliary coupler solvents*

Coupler solvent	λmax	Dmax
DNP*	670	2.94
TB*	671	2.34
TL*	657	2.78
DCP*	667	2.50
Sample 1	667	1.91
Sample 2	664	2.82
Sample 3	658	2.86
Sample 4	662	2.96
Sample 5	661	2.76
Sample 6	665	2.68
Sample 7	664	2.68
Sample 8	665	2.87

* DNP Di-n-butyl phthalate; TB tributyrin; TL trilaurin; DCP - dicyclohexyl phthalate

Figure 4 *The effect of coupler solvent on λ_{max} and D_{max}*

an aqueous phase containing gelatin and the dispersing agent ALKANOL XC (Dupont) in a 10:1 weight ratio. Each of the resulting mixtures was passed through a colloid mill to disperse the coupler-containing oil phase in the aqueous phase as small particles ... The typical resulting comparative dispersions contained 1.50% by weight of coupler, 1.50% by weight of dibutylsebacate, 4.50% by weight of ethyl acetate, 6.0% by weight of gelatin and 0.60% by weight of Alkanol XC ...

Clearly the coupler solvent, the surfactant, the coupler and indeed the ratios may change, but the method of preparing the dispersion will be similar – at least for conventional couplers.

Coupler solvents have also been studied for their effects on physical parameters. In their 1971 patent, the workers at the Minnesota Mining and Manufacturing Company introduced a range of new coupler solvents,[6] Figure 5.

They claimed that

... The dispersions of the colour formers obtained by the crystalloidal solvents of the present invention, show physical characteristics such as a drop in the average diameter of the dispersed phase and noticeably improved distribution curve of the phase itself, with respect to those ones obtained with already known compounds. In particular a smaller average diameter and a more narrow distribution curve are obtained. This implies an increase in the dispersion's covering power and allows an improvement in the colour quality and in the definition of the colour materials. Besides, the solvents of the present invention, not only exhibit a high dissolving power with regard to the

N-amylformamide
N-hexylformamide
N-propyl-N-butylformamide
N-octylformamide
N,N-dibutylformamide
N,N-dihexylformamide
N,N-di-2-ethylhexylformamide
N,N-dioctylformamide
N-ethyl-N-cetylformamide
N-benzylformamide
N-(p-methyl) benzylformamide
N-(p-methoxy-benzyl)formamide
N-benzyl-N-butylformamide
N-(p-methyl)benzyl-N-butylformamide
N-(p-methoxy)-benzyl-N-butylformide
N-formyl-N-(N',N'-diethyl)-acetamidoformamide
N-phenyl-N-butylformamide
N-hexadecyl-N--carboethoxy)-phenylformamide
N-(p-methyl)benzyl-N-carboethoxymethylformamide
N,N-diphenylformamide
N,N-dibenzylformamide
N-butyl-N-(m-methyl)-phenylformamide

Figure 5 *Some modern coupler solvents introduced by 3M*

Material	Scratchability expressed in grams
A	320
B	360
C	385
D	600
E	110
F	110
G	215
H	300

Figure 6 *Scratchability results using modern coupler solvents*

common dispersing adjuvants, but, surprisingly they perform a specific action with regard to the gelatin layer containing them, conducive to a lesser scratchability of the layer itself, which can be seen, in particular, on the material after processing ...

The same dried samples were subjected to scratchability tests as described by Sheppard and Schmitt. Briefly this method consists of making a single scratch on the surface of the photographic film under a load increasing at a definite rate and measuring the minimum load sufficient to scratch the said surface. The scratchability is expressed in grams and it decreases with the increase of the number of grams. The results obtained in the study mentioned above are recorded here in Figure 6.

The samples A through H are different coatings of the same materials, coated with different coupler solvents. There appear to be clear advantages (lower scratchability numbers) with some of the samples, and therefore the coupler solvents, relative to others.

7.1 The Preparation of Polymeric Coupler Dispersions

Chen *et al.*[7] preface their discussion concerning the preparation of dispersions containing polymeric couplers in the following way

... One difficulty encountered with the use of polymeric couplers in photographic materials is in providing good dispersions of the polymers, which allow incorporation of the polymers in the photographic material. Generally, three methods have been employed in the past for dispersing polymeric couplers.

In the first method, the polymeric coupler is formed by solution polymerization and is isolated by precipitation from a poor solvent. As is known in the polymer art, solution polymerization employs a solvent as the reaction medium in a homogeneous system, [Billmeyer, Jr., Textbook of Polymer Science, Wiley-Interscience (1971)]. The resulting solid polymer coupler is then dissolved in ethyl acetate, with or without additional coupler solvents, mixed with gelatin and surfactants, and passed through a colloid mill to produce a fine dispersion. However, this method is disadvantageous in that it requires many energy-consuming steps.

A second method for dispersing polymer couplers requires the incorporation of a large amount, usually greater than about 40 weight percent, of ionic monomers in the polymeric coupler. The resulting polymers are water-soluble and can be directly mixed with gelatin and coated with a silver halide emulsion on a support. However, this method is disadvantageous in that the use of the relatively large amount of ionic monomers increases the equivalent weight of the polymers, results in gel-polymer interactions and causes increased wandering.

A third method for dispersing polymeric couplers comprises the formation of water-dispersible polymers by emulsion polymerization or suspension polymerization as taught, for example, in the Monbaliu et al U.S. Pat. No. 3,926,436, the Van Paesschen et al U.S. Pat. No. 4,080,211, the Yagihara et al U.S. Pat. No. 4,474,870, British Reference No. 2,092,573 and European Pat. Application No. 321,399.

The method employing emulsion polymerization is usually preferred. However, such methods are disadvantageous in several respects. That is, owing to the low solubility of many coupler monomers in water, organic solvents are required in the emulsion polymerization. This tends to reduce the stability of the resulting polymer latexes and to reduce the percentage of solids in the polymer products. Additionally, the compositions of the polymeric couplers made by emulsion polymerization are difficult to control because of the heterogeneous nature of the system. Coupler monomers usually are solid and owing to their low water solubility, they tend to precipitate out in the aqueous phase and fail to copolymerize with other comonomers. Finally, polymeric couplers made by emulsion polymerization methods can not be isolated and redispersed in water...

7.2 Dye Cloud Formation

Silver laydown, or more correctly, the number of available imaging sites, in combination with the coupler, that may be either 2-equivalent or 4-equivalent with respect to silver usage, are large factors in the granularity of the final image. It is possible to create films with the equivalent silver laydown in gm m^{-2} but with different numbers of silver halide crystals. The number of imaging sites is therefore of interest. Silver halide crystals could either be of different sizes, different shapes (t-grains *vs*. cube for example) or a combination of the two.

The mechanism of dye cloud formation from a latent image has been the subject of much research. Tong produced one of the classic texts on the subject in his 'Mechanism of dye formation and related reactions'.[8] Tong commented (p. 339)

> *... Colour development begins with the oxidation of the p-phenylenediamine to quinonediimine (QDI), This is usually the slowest sequence in the dye forming process ...*

He later commented (p. 351)

> *... the image structure of colour photographic materials is determined by the distribution of dyes around each grain, the latter is referred to as a 'dye cloud' ... The dye cloud is a result of relative reactions involving QDI such as coupling and non dye-forming competing reactions as well as diffusion ... The source of the QDI is located at the grain surface ... It is assumed that dye is deposited where it is formed ...*

The result of these various competing reactions is the formation of dye clouds. The dye cloud aggregates can be measured at the macro density level under an optical microscope using essentially a dye cloud counter. This physical measurement is called 'granularity'.

But what is granularity?

Granularity is often defined as the objective measurement of graininess, where graininess is the visual sensation associated with non-uniformity in a photographic image. For example the picture below shows a typical country scene. It has been deliberately chosen so that there is a neutral grey in the foreground, which has dappled sunlight falling on it, Figure 7.

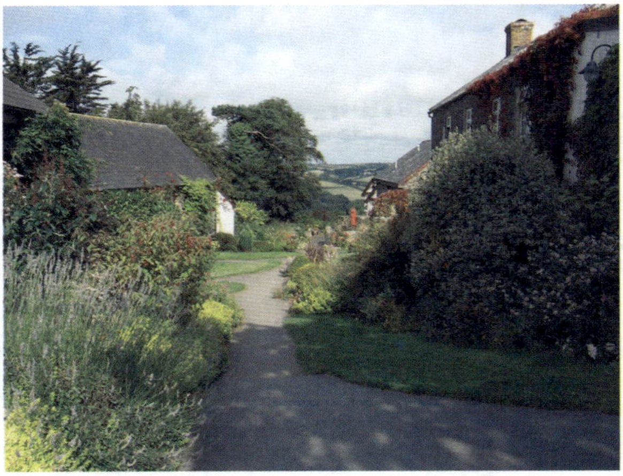

Figure 7 *Typical country scene*

While in an actual print the granularity would be random, Figure 8 is a modified image where the granularity has been uniformly increased by 10% to demonstrate the principle.

The reader should see some dye clouds in the sky and shadow on the path. As an extreme example the granularity has been increased by a further 40%, taking the overall increase to 50%. In the picture below, Figure 9, the path has been magnified to demonstrate the effect in shadows and daylight. Note that the graininess/granularity has once again been uniformly increased. This would not be the case in a real colour print as one dye would undoubtedly increase more than another.

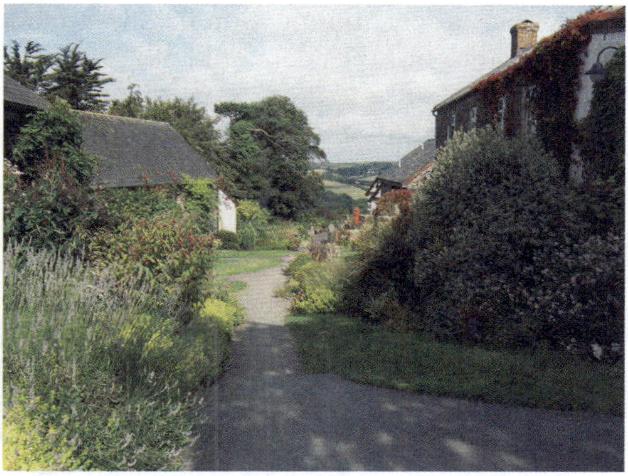

Figure 8 *Uniform increase in granularity*

Figure 9 *Magnified part of the scene with a 50% increase in granularity*

In all three cases the overall dye density has remained unchanged, it is only the degree of dye cloud aggregation that has been changed. At the same silver weight and silver grain crystal structure, or 'morphology' as it is known, the effect is related to the number of silver equivalents which are used in dye formation which is related to the coupling-off group from the coupler molecule.

The following diagram uses one colour and regular shapes to demonstrate the more complex scenario of a normal image, Figure 10.

More dye is formed for the same silver weight, or less silver can be used for the same dye density using 2-equivalent couplers, compared with 4-equivalent couplers. Silver halide inter-grain distance is also an issue, as are the relative amounts of silver halide and coupler.

In all of the examples above, no mention has been made of the light scatter by the silver grains, nor the effect of light absorption on the layer below. Figure 11 is part of an actual cross-section of a commercial film. The layers that are displayed are the yellow filter layer, the fast magenta dye-forming layer and the slow magenta dye-forming layer. A complete

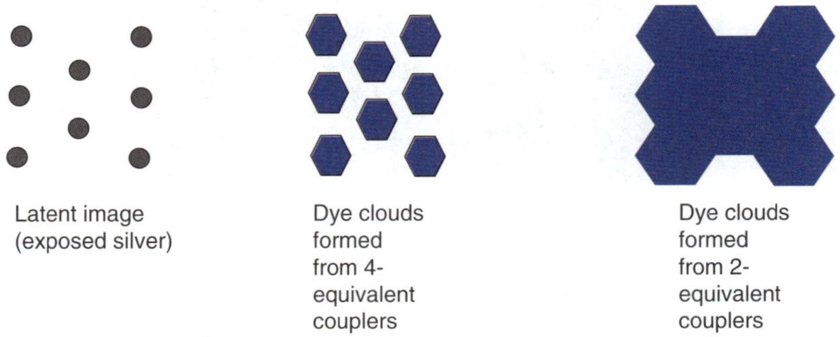

Figure 10 *The relationship between dye clouds, silver image and coupler equivalency*

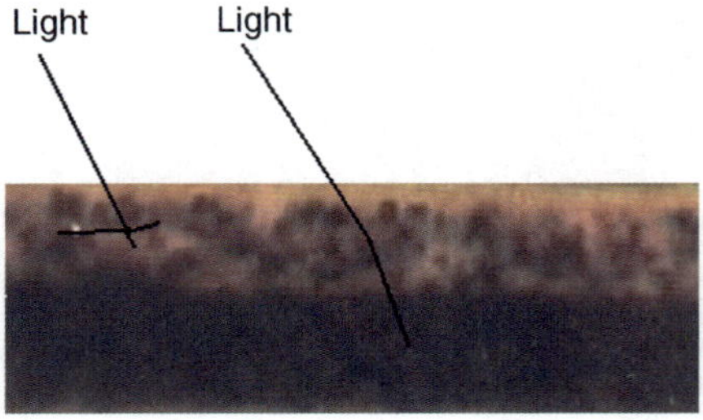

Figure 11 *Cross-section of an actual film*

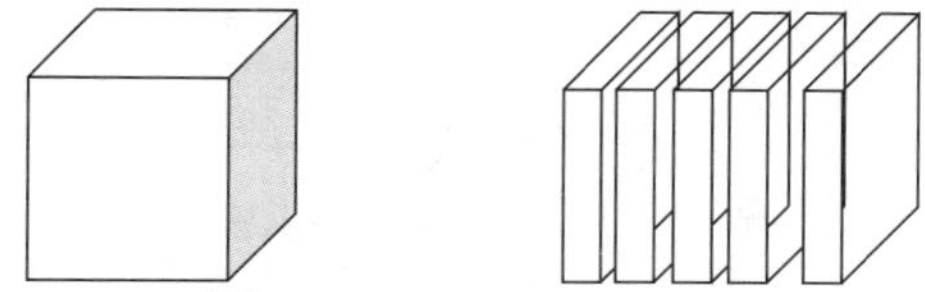

Figure 12 *The concept of increasing the number of silver halide crystals for the same weight of material*

description of the various layers in colour negative films can be found in Chapter 9. In this case the cross-section is of unprocessed film.

On the left side, light enters the upper layer and has been scattered by the silver halide grains to such an extent that none of the light has been absorbed by the layer below. On the right side, the light has experienced no effect from the above layer. Fortunately, there is a mechanism for maintaining the desired granularity at the same time as reducing, and therefore minimising, light scatter. The mechanism used is to change the shape of silver halide grain in one dimension using the same silver halide weight, which increases the number of potential imaging centres. With an increase in the number of silver halide grains, the film builder has the option of either reducing the granularity of the layer, or reducing the silver halide laydown level. A reduction in silver halide laydown rate will potentially decrease the likelihood of light scatter by the silver halide grains as there will be fewer of them.

Consider a cubic silver halide grain, Figure 12.

In practice, of course, it was not practical to suggest that each silver halide crystal is cut into many thinner crystals. The scientists working on silver halide crystals determined a means of producing these types of grains through concentration, temperature, vAg and stirrer control.

Figure 13 shows a micrograph of a typical example of this type of emulsion.

The method of precipitation of these tabular grains did, however, offer an improved granularity at the same silver laydown rate, with the potential of reducing the silver weight if granularity is not an issue.

7.3 Dye Stability

Most dyes are subject to the conditions surrounding their storage, typically dyes can be affected by

- heat

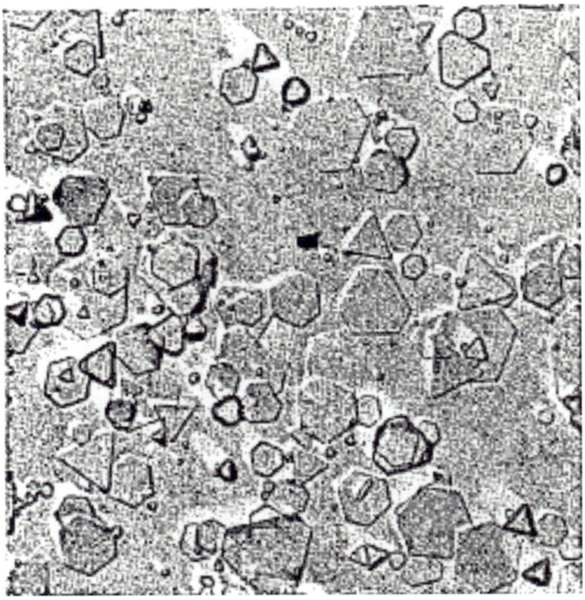

Figure 13 *A typical example of silver halide crystals*

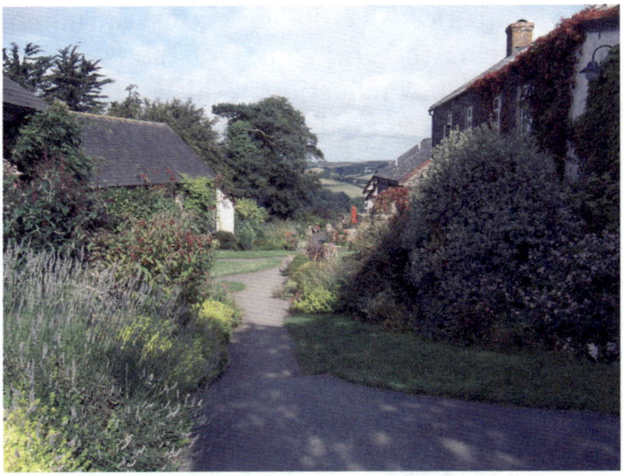

Figure 14 *Simulated dye fade*

- humidity
- light
- a combination of the above
- glue from storage boxes
- atmospheric oxygen
- peroxides from paints
- *etc.*

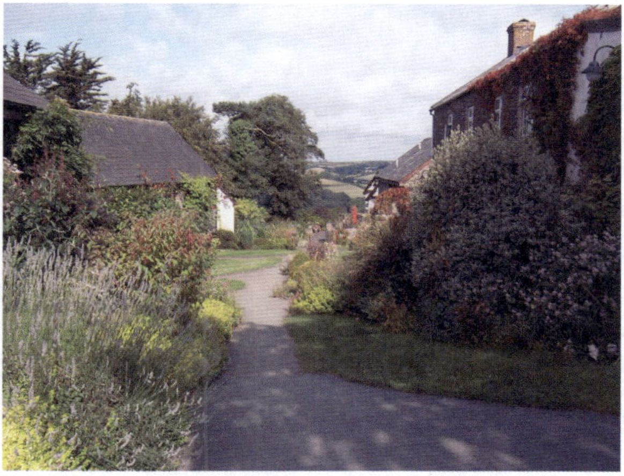

Figure 15 *Simulated dye fade*

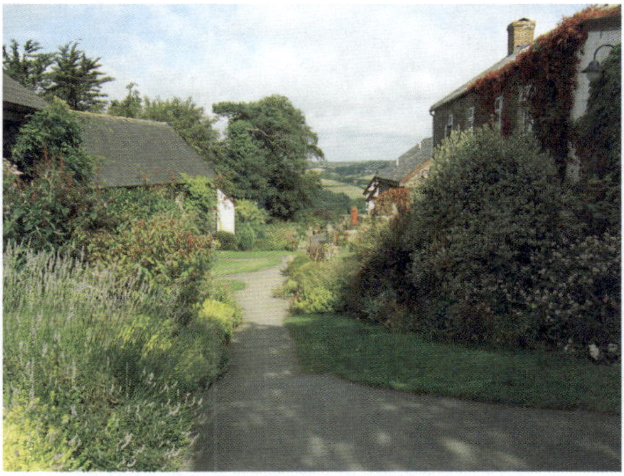

Figure 16 *Simulated dye fade*

Different dyes are affected to different extents. The following three pictures simulate the effect of dye fade on a colour print. In each case a different dye has been affected, while the other two dyes remain unaffected, Figures 14, 15 and 16.

Perhaps the most common dye fade that has been observed by the average consumer reflection print, would be the magenta dye fade, Figure 15. These simulated reflection prints show a single dye faded, whereas in practice all three dyes will fade. It becomes a matter of the

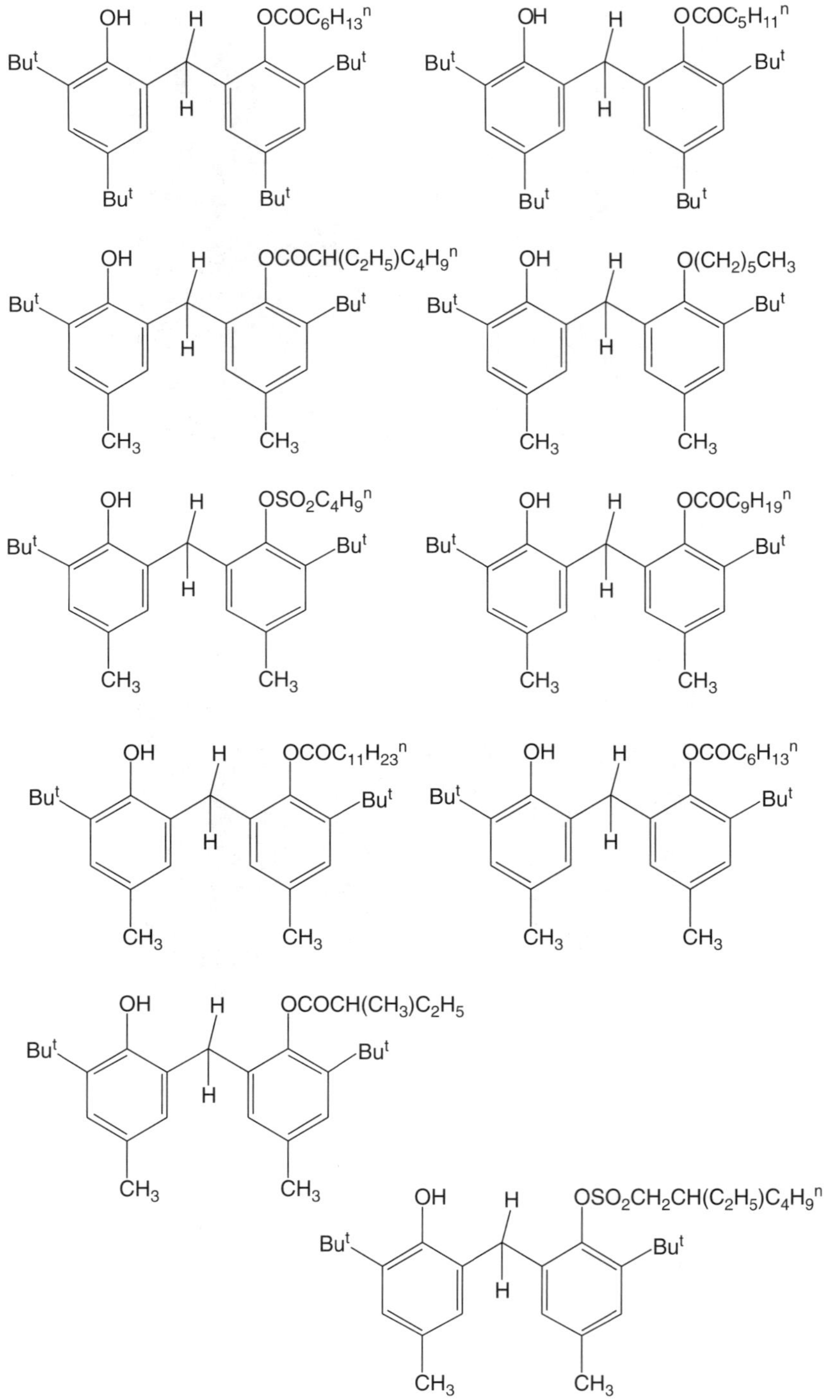

Figure 17 *Potential dye stabilisers*

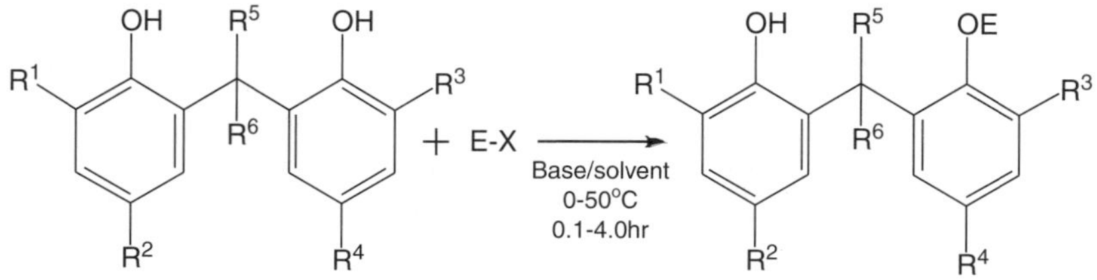

Figure 18 *A general route to dye stabilisers*

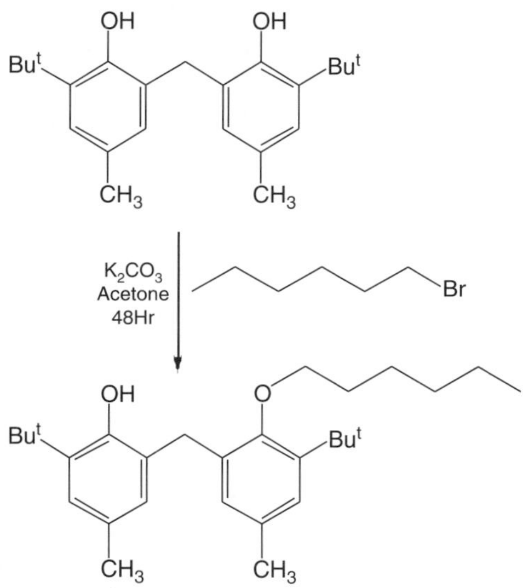

Figure 19 *A route to a dye stabiliser*

relative kinetics of dye fade, which varies between the photographic manufacturers, storage conditions and indeed the dyes themselves.

Correlations between structure and dye fade are routine during the research and development of new photographic products. In most cases a coupler is selected from a range of variants, the dye fade for which is acceptable for the proposed application. In some cases, however, dye stabilisers are added to the film or paper product formulation.

Should a stabiliser be required the preferred method of adding a dye stabiliser to a product is to co-disperse the stabiliser with the coupler. Krishnamurthy and Jain[9] proposed a class of stabilisers of particular benefit to yellow dyes in their European patent, Figure 17.

There are three potential syntheses of these stabilisers, which are described in Figures 18, 19 and 20.Where E-X are electrophiles, such as

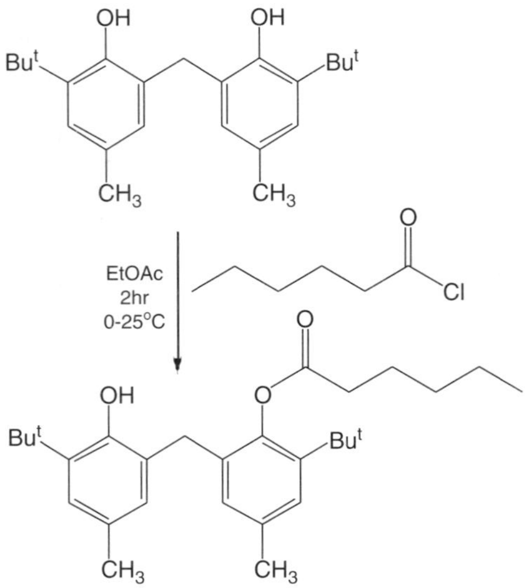

Figure 20 *An alternative route to a dye stabiliser*

acylating agents, aroylating agents, sulfonylating agents and phosphorylating agents.

References

1. D. Clarke, L.J. Leyshon and K.E. Smith, EP 1,197,798, Eastman Kodak Co.
2. P.B. Merkel and J.N. Poslusny, US 6,174,662, Eastman Kodak Co.
3. E.E. Jelley and P. Vittum, US 2,322,027, Eastman Kodak Co.
4. F.C. McCrossen and H.J. Osborn, US 3,779,765, Eastman Kodak Co.
5. J.N. Poslusny, P.B. Merke, D.A. Steele, D.M. Michno, D.S. Ross, R.K. Rothrock, P.L. Zengerle, US 6,680,165, Eastman Kodak Co.
6. Minnesota Mining and Manufacturing Company, GB 1,357,372.
7. Tien-Teh Chen, S.W. Cowan and E. Schofield, US 5,455,147, Eastman Kodak Co.
8. T.H. James, *The Theory of the Photographic Process*, 4th edn, Section II by L.K.J. Tong, MacMillan Publishing Co. Inc., New York, 1977, ISBN 0-02-360190-6.
9. S. Krishnamurthy and R. Jain, EP 0 632 324, Eastman Kodak Co.

CHAPTER 8

The Chemistry of Colour

Arguably, the seminal text on the reproduction of colour was written by R.W.G. Hunt.[1] Hunt's text covers all aspects of colour reproduction from all perspectives. The short discussion within this chapter is concerned with the chemistry of colour reproduction within the photographic system. The interested reader should consult the above reference for a more detailed account of all the many other aspects of colour science.

Dye density absorption curves for three typical dyes used in colour negative film are shown as Figure 1. In all three cases the dyes show unwanted absorptions. Dyes can be synthesised that have a much reduced half-band width. Unfortunately the use of these dyes can lead to other issues relating to the faithful reproduction of the original scene were they to be used in the construction of a film.

The curve depicted in black in Figure 1 is the integration of all three of the dye density spectra. It is not a flat line of equal dye density across all of the wavelengths, so there are colour issues that present themselves as challenges to the film builders.

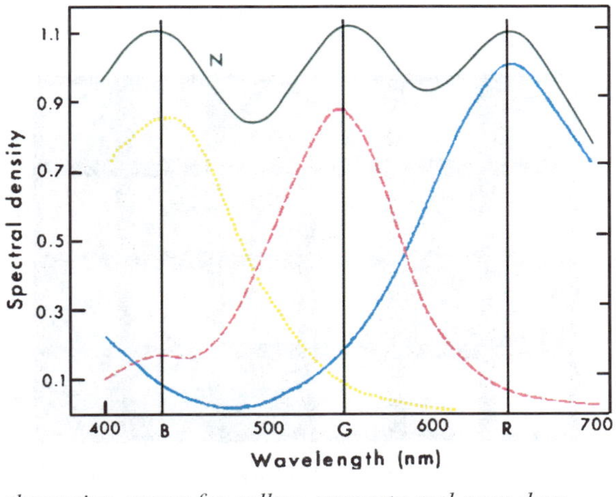

Figure 1 *Dye absorption curves for yellow, magenta and cyan dyes*

While Chapter 9 and the subsequent chapters will consider the techniques that are used to construct various films and papers, it is worthwhile showing here a curve of the photographic response of a light sensitive product to visible light. Figure 2 depicts the photographic response as measured in dye density to the logarithm of the exposure. This particular plot has been shown to be the most practical method of viewing this data and is a reproduction of Figure 17 – see Chapter 1 for more details concerning the generation of this plot.

The crucial issue is that the photographic response, *i.e.* density, to exposure produces an equal colour balance throughout the exposure range. There are several methods open to the film builder, which enable a validation that the film has faithfully recorded the colours in the original scene. The most useful chart, which is used extensively in the photographic trade, is known as the 'Macbeth checker chart' an example of which is shown as Figure 3.

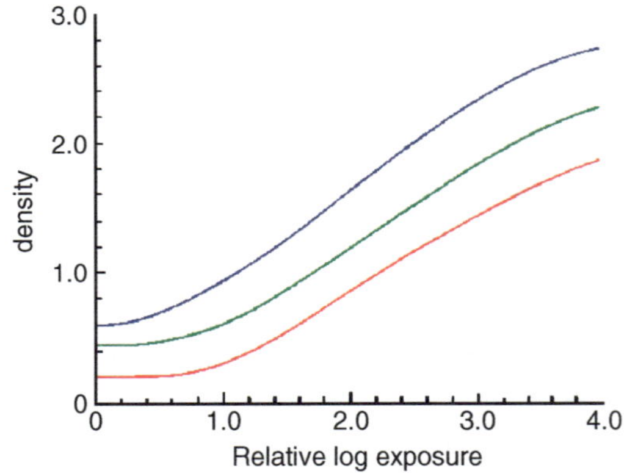

Figure 2 *Typical colour negative film density v log exposure curves*

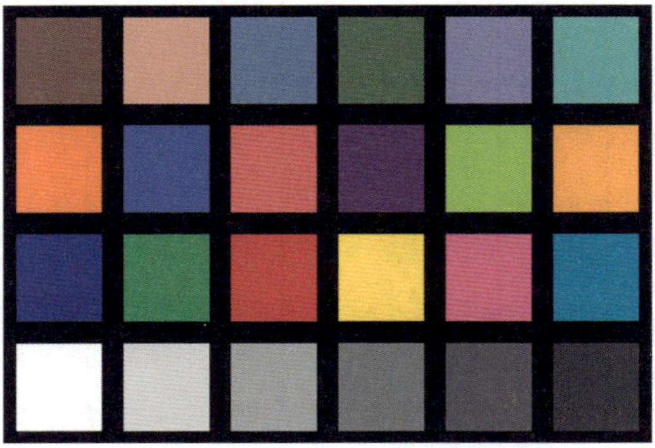

Figure 3 *A typical Macbeth colour checker chart*
(Reproduced with permission from X-rite)

Each one of the patches on the chart represents a colour that is mandated to be as accurate as possible if one is designing a film to be a faithful reproduction of an original scene. Some film manufacturers produce films which, when printed onto photographic paper, show greener greens and bluer blues. One reason for this is that holiday picture takers can look back on their holiday and be pleased that the sea was indeed blue and the grass was green – perhaps more than they realised! In practice, there is still one patch that will need to be accurately reproduced and that is on the top row second from the left – at least for white Europeans. This patch is the 'flesh' patch. No matter what the colour rendition of blues and greens, if facial colours deviate from a faithful reproduction of the original there will be problems accepting the final pictures. Flesh to neutral is always of paramount importance, especially so if the reproduced flesh patch were to have a green or blue tinge!

One of the issues faced by the technical communities of the manufacturing divisions is therefore the relative distance of one curve to each other in Figure 2. If all of the curves run 'high' or 'low' at the same time during manufacture, then the film will reproduce colours effectively (there are nevertheless limits). If on the other hand, one of the curves runs 'high' while the others are running 'low', or indeed on aim, the colour balance will be severely affected and depending on the severity it may result in the film being scrapped, Figure 4.

The same argument is valid for transparency materials and for colour paper products. Once manufactured, there is nothing that can be done to correct one colour record, without causing issues for the other colour records. This phenomenon is therefore of paramount interest to the manufacturing community during manufacture of a colour product.

Figure 1 showed the dye absorption curves of the cyan, magenta and yellow dyes produced by coupling oxidised colour developer with the

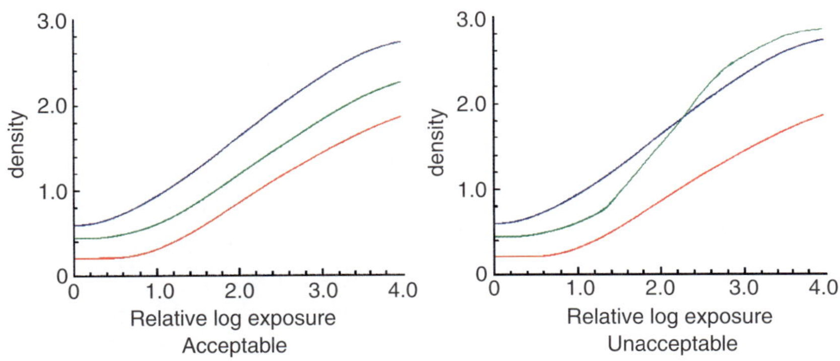

Figure 4 *Examples of acceptable and unacceptable photographic response*

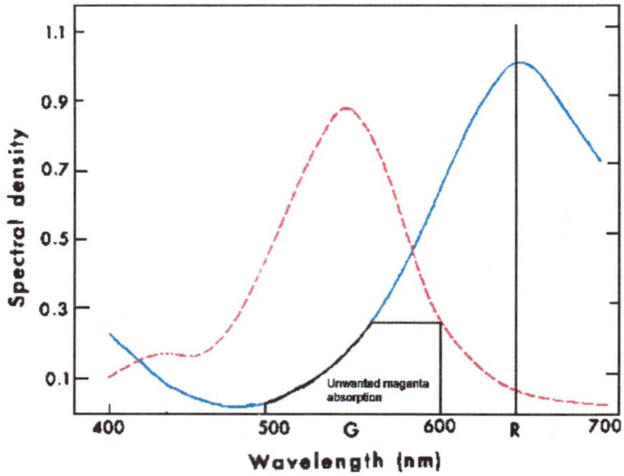

Figure 5 *Unwanted magenta dye absorption from the cyan dye*

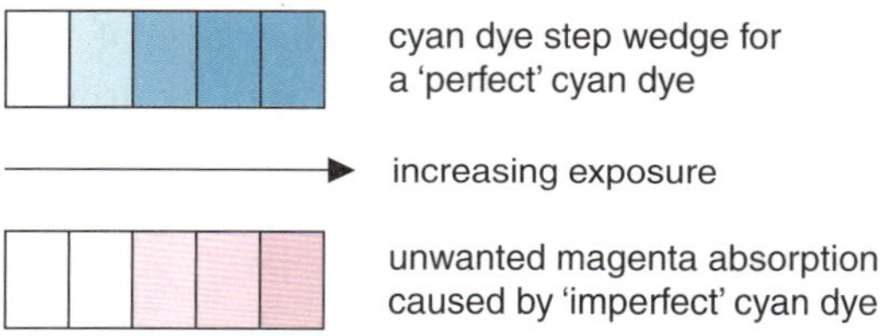

Figure 6 *Unwanted absorptions shown as a step wedge*

relevant coupler. Figure 5 shows two of these curves, namely the cyan and magenta, on the same plot.

There is a region of the absorption spectra between 500 and 600 nm where the cyan dye is contributing to the magenta absorption. There is a similar but reciprocal contribution from the magenta dye in the cyan dye region. Let us consider the consequences of one of these unwanted absorptions, and concentrate on the unwanted magenta absorption of the cyan dye. Figure 6 shows incremental steps of cyan dye, which for the purposes of this argument is assumed to be a 'pure' dye with no magenta absorptions. Underneath this step wedge is the corresponding magenta curve caused by the unwanted dye absorption from an imperfect dye.

Figures 6, 7, 8 and 9 simulate the issues concerning the unwanted absorptions using a five exposure step wedge. A test object such as this with just five steps would not be used in practice, more likely one of 21. However, this wedge adequately simulates the issues.

The magenta step wedge has been exaggerated for the purposes of argument so that the effect can be seen more clearly. In this case, the solution to the problem of cyan dye colour fidelity is to incorporate a coupler which is magenta coloured and which, when reacted with oxidised colour developer, produces a cyan dye, Figure 7.

Once again the effect has been greatly exaggerated for demonstration purposes. A conventional dispersion is made in the normal manner from these coloured couplers – see Chapter 7 for a discussion of dispersion formation. A small amount of this coupler dispersion is added to the cyan dye-forming layer, in addition to the colourless cyan dye-forming coupler previously discussed – Chapters 6 and 7. The ratio of the two couplers is of the order of 1:10 coloured to colourless coupler, but varies between the manufacturers, as the amount of unwanted dye absorption is different for the different dyes.

The combined effect of these two magenta absorptions, Figure 8, is a constant exposure of magenta dye. This additional and constant dye density can be allowed for in the printer exposure settings during the printing of the negative onto the photographic paper. Most of the manufacturers have slightly different minimum densities compared with each other, and sometimes within the films of a given film family. Printing dyes are therefore included in the film design, the amounts of which vary from film to film so that the printing density for a specific test object remains constant within a film family.

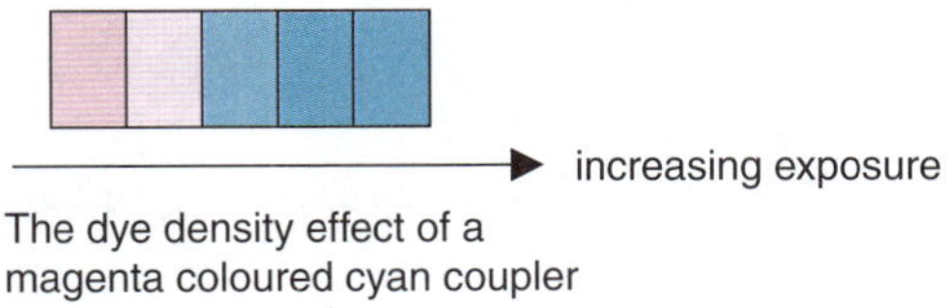

Figure 7 *A step wedge of a magenta coloured cyan coupler*

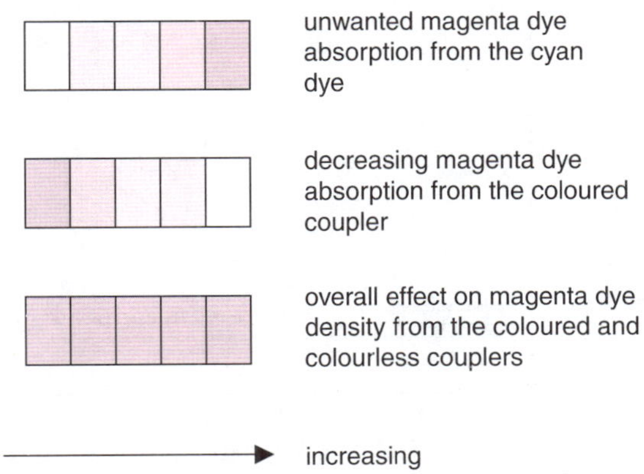

Figure 8 *Colour correction of unwanted magenta dye*

This type of colour correction is used extensively in all colour negative films, from all manufacturers. Indeed it is the effect of the coloured couplers that provide film negatives with some of their distinctive colour, often observed near the perforations, because the coloured dyes have not reacted with oxidised colour developer outside the image areas, *i.e.* near the perforations, and so the colour is a mixture of these dyes.

In practice most colour negative films only use two coloured couplers, sometimes called 'masking couplers'. The two most common masking couplers are a yellow coloured magenta dye forming coupler, and a magenta coloured cyan dye forming coupler, Figure 9.

A magenta coloured cyan coupler has the density log exposure profile of Figure 10, where red and green have been used to replace cyan and magenta, respectively, as these colours are the ones that the mask is intended to affect.

The generation of these types of masking couplers is relatively straight forward, in that it is the coupling-off group that is responsible for the initial colour. This coupling-off group has no ballast and therefore washes into the processing solutions upon processing. Care must be

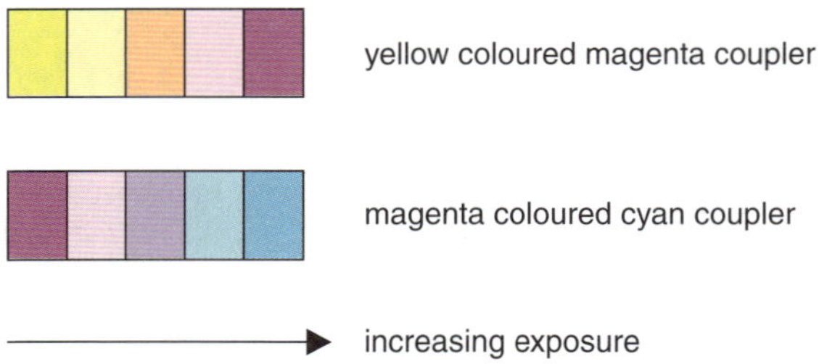

Figure 9 *Step wedges of the common colour correction couplers*

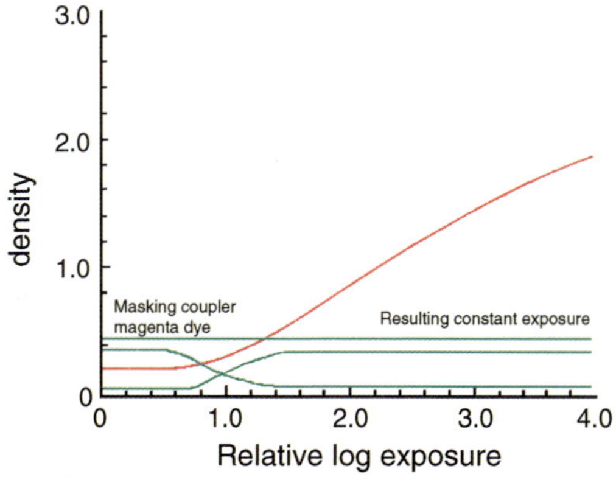

Figure 10 *Exposure curves representing colour correction*

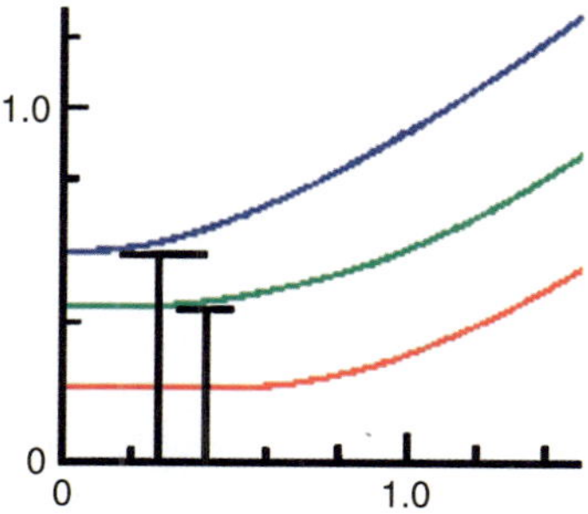

Figure 11 *The effect of printing dye density on the minimum densities of each colour record*

taken during the design of this coupling-off group that the resultant dye will not cause dye stain, when the same solutions are used to process the next film. These dyes are therefore also chosen so that they react with the sulfite or other chemicals in the processing solutions, and produce colourless bi-products.

The colours associated with processed colour negative films in the unexposed areas of the negative are therefore a mixture of the yellow and magenta dyes. As will be seen later, there is a yellow coloured layer in a colour negative film, but the colour is removed during processing and so will not be seen in the final processed film negative. These hues manifest themselves on the density/log exposure curve, Figure 2, as increased minimum density at zero exposure, Figure 11.

In this particular example of a colour negative film, the red curve has no contribution to the density from a masking coupler. Both the green and blue curves do have masking couplers coated in the relevant layers, which adds to this minimum density. Additionally, the yellow layer has a contribution from a yellow filter layer, which will be described in more detail in Chapter 9. There will be fog, or dye density, produced in non-image areas, for each of the colour records. All of the minimum densities are affected by the printing dyes mentioned above.

In general masking couplers are prepared from the reaction of the parent 2-equivalent coupler with a diazonium salt, see for example ref 2, Figure 12 or ref 3.

In this case, the coupling-off group is the 4-hydroxyphenylazo group, which will be yellow. The parent coupler is a pyrazolone, which will form a magenta dye with oxidised colour developer. The example given, Figure 12, is but one of many examples that are covered by the aforementioned patent. Clearly any aniline capable of forming a diazonium salt can be reacted with any coupler to produce a vast range of masking couplers.

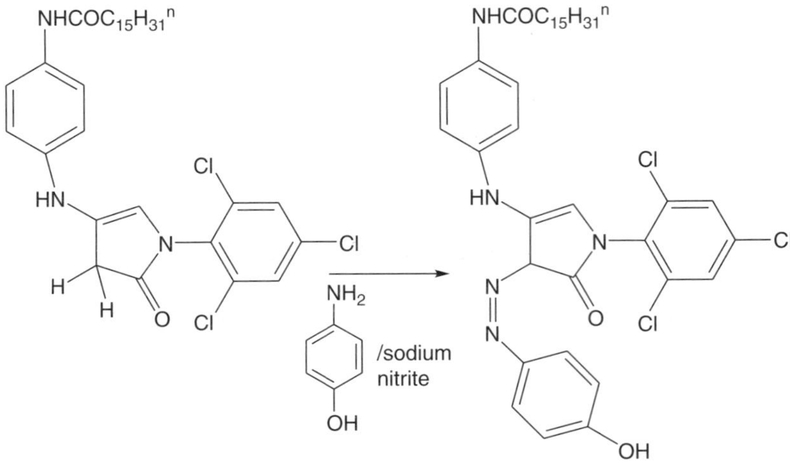

Figure 12 *The synthesis of a masking coupler*

Magenta coloured cyan couplers could also be produced by the same synthetic route, however in practice, a more challenging synthetic route has been adopted. Figure 13 details a synthesis of a representative sample of an Eastman Kodak Co. magenta coloured cyan coupler.[4]

8.1 Inter-Layer Inter-Image Effects (IIE)

In his book *The Reproduction of Colour*, p. 280, Hunt[1] defines inter-image effects as

> *... the effects that may be present whenever different development rates occur in adjacent layers. This can happen in several ways:*
>
> - *as a developer penetrates a multi-layer colour material, it will normally be partially exhausted by the time it reaches the bottom layer, hence if the development is not carried to completion in all layers, it may be necessary to make the bottom layer faster or higher contrast*
> - *they can also be caused by the degree of development in one layer being affected by the release from a neighbouring layer of development-inhibiting agents*
> - *a third possibility is that of oxidised colour developer wandering from one layer to another ...*

Inter-layer inter-image effects (IIE) can be measured by using multiple exposures. Figure 14 provides a typical example of one of the many exposures that might be used to measure IIE.

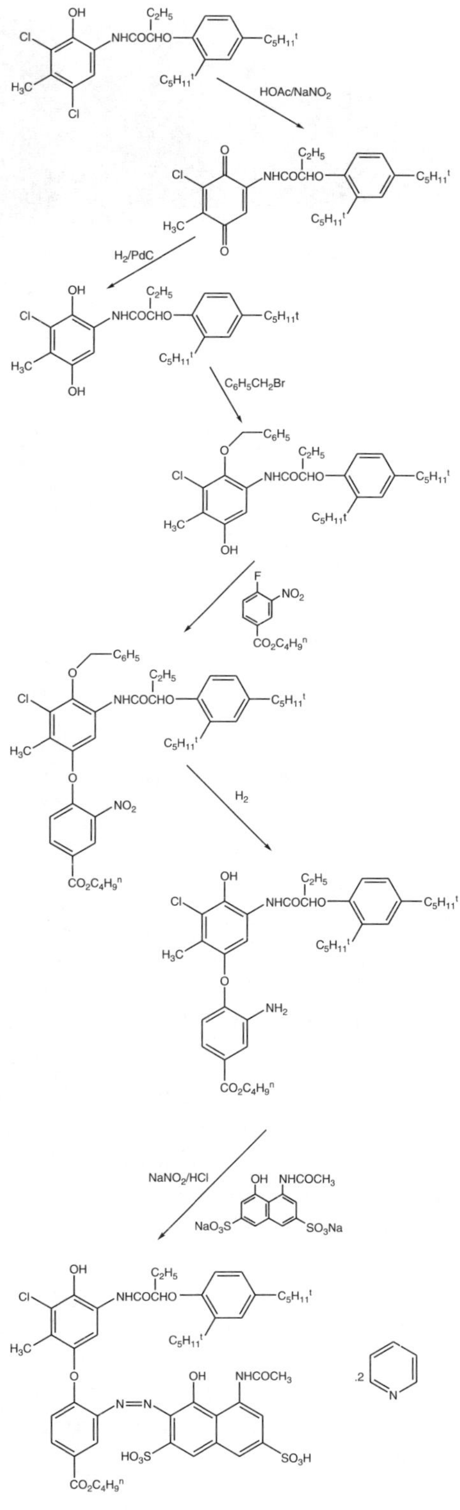

Figure 13 *The synthesis of a magenta coloured cyan coupler*

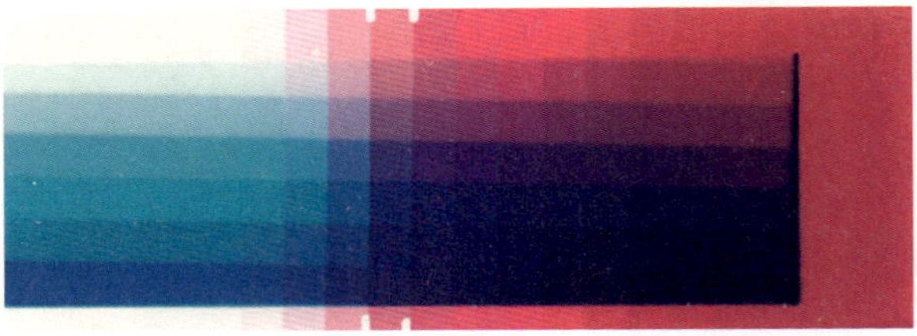

Figure 14 *A multiple exposure used to evaluate interlayer inter-image effects*

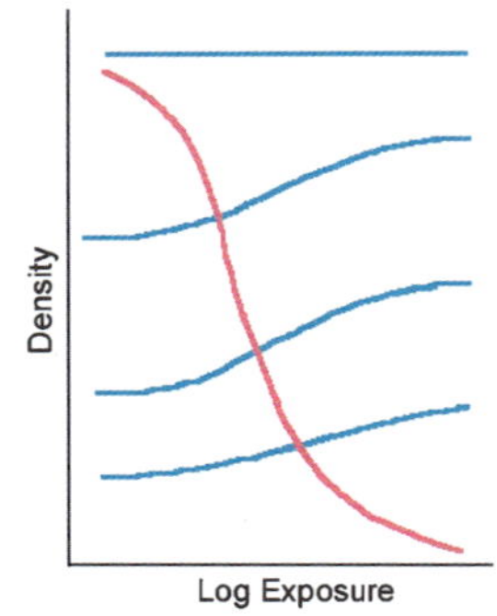

Figure 15 *A plot of interlayer inter-image effects*

In this particular colour paper example, a cyan wedge has been exposed orthogonal to a standard magenta wedge. There are many such exposures where the standard wedge can be yellow, magenta, cyan, red, green or blue. With the exception of the colour of the main wedge and of the other colour that has been used, any of the remaining colours can be flashed through the orthogonal wedge. IIE plots can then be constructed by reading the grid, an example of which is provided as Figure 15.

From this and the other plots, the total inter-image effects can be determined and remedial action implemented, depending on the extent and cause of the issues. For example, if there is a developer access issue perhaps the hardener level of the paper can be altered so that the gel is allowed to swell to a greater extent during the development phase. The so-called 'gel to junk' can be altered, *i.e.* the inter-grain distance, thereby allowing developer access.

8.2 Development Inhibitor Couplers

The second of Hunt's IIE causes relates to the use of development inhibitor chemistry. These chemicals are added to multilayer colour films to control sharpness.

The sharpness of a film can be measured by a test object, which has been specifically designed for the process. Consider an approximation of an actual test object, which is representative of the issues that are prevalent in the design of a film. In an actual test object the area occupied by the lines will be constant and there will be more lines per mm for the thinner lines. In this particular representation, the number of lines has been kept constant, but the line thickness and spacing reduced, to demonstrate the principle that there comes a point when the eye can no longer resolve differences between the lines – even though there are discrete lines, Figure 16.

The film sharpness is governed by chemistry and there will be a point where no further increases in sharpness will be detected by the film. In the case of a film, the dye clouds formed from latent images will slowly become merged, as there will be less definition to the discrete line. Figure 17 represents lines of narrow spacing and has been magnified to demonstrate the principle.

In practice, of course, it is more likely that the average photographer will look at a picture and not a test object. Reproduced below are two pictures of part of the country scene that has been used previously shown here to demonstrate the effect of a sharp and unsharp picture, Figures 18 and 19, respectively. This photographic effect is controlled by the use of chemicals called development inhibitor releasing couplers (DIRs) or development anchimerically releasing couplers (DI(A)Rs).

DIR and DI(A)R couplers produce sharp edges on a photographic image by preventing development in the unexposed regions of either black and white or colour materials, Figure 20.

Figure 20 represents an X-ray line exposure of a single layer, where the trace shows the effect of image prevention by development inhibitors. The effect of this selective 'removal' of the image actually provides

Figure 16 *A representation of a test object used to measure film sharpness*

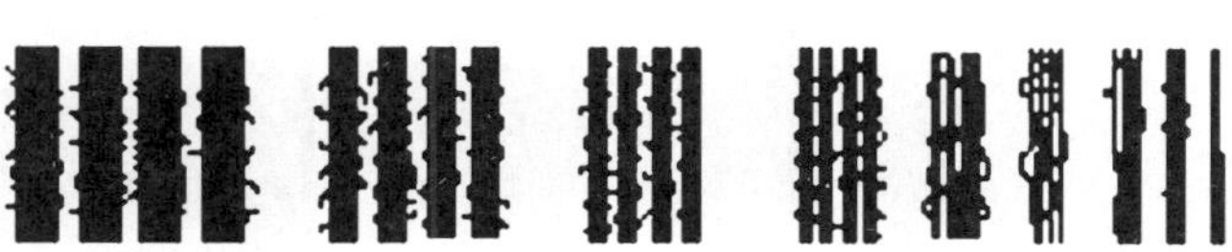

Figure 17 *Magnified effects of exposing the sharpness test object onto the film*

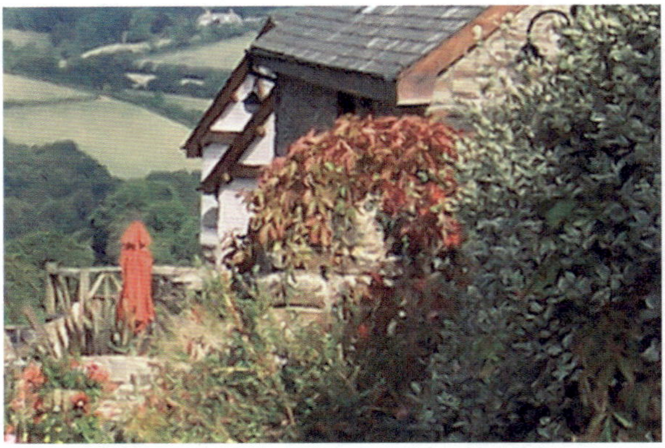

Figure 18 *A sharp area of the country scene*

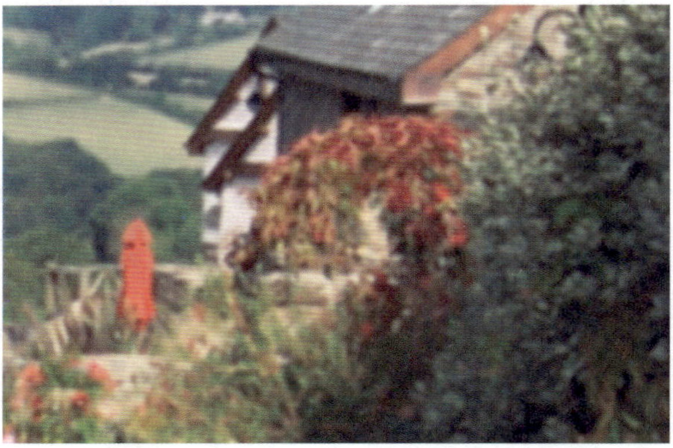

Figure 19 *An unsharp area of the country scene*

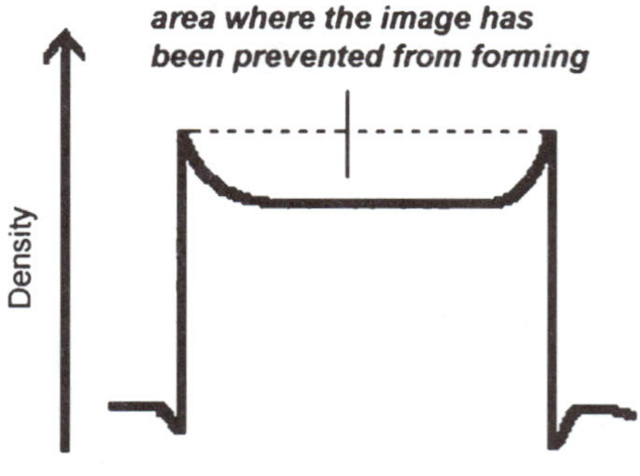

Figure 20 *The micro level effect of development inhibitor chemistry*

sharper edges for the remaining image. Figure 21 shows the effect of the selective 'removal' of image at the macro level.

DIR and DI(A)R couplers each release development inhibitors from the coupling position during coupling with oxidised colour developer. In most cases, the coupling-off group is a sulfur containing moiety. The difference between a DIR and a DI(A)R is that, in the case of a DI(A)R, the inhibitor fragment is released from the coupling position of the coupler attached to a switch. This switch then releases the mercapto (or other development inhibitor fragment) inhibitor fragment after a period of time, Figure 22.

The switch/inhibitor fragment allows for migration of the inhibitor fragment to a layer that would be inaccessible for the inhibitor fragment on its own, Figure 23.

The ability to affect image development some distance away from the latent image has many advantages for the control of sharpness. Both DIR and DI(A)R couplers have been synthesised for all three colour records. Additionally, the ability to reduce the contrast of an adjacent layer, Figure 21, also has implications in the control of colour rendition.

Development inhibitor couplers are used at approximately 5–10% of the laydown rate of the respective imaging coupler. The dye produced from them must therefore be of the correct hue and stability. Furthermore, the

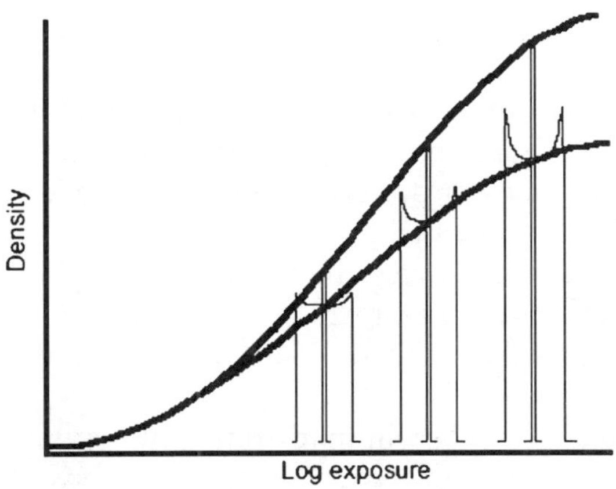

Figure 21 *The macro level effect of development inhibitor chemistry*

Coupler – Inhibitor   coupler – switch – inhibitor

DIR   DI(A)R

Figure 22 *The difference between DIR and DI(A)R couplers*

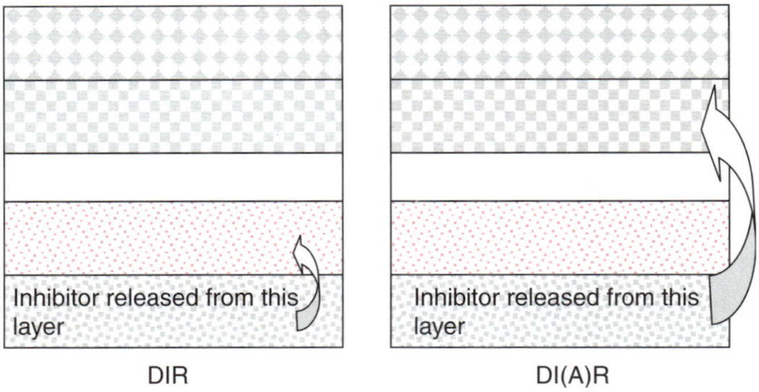

Figure 23 *DI(A)R fragments can travel further than DIR fragments*

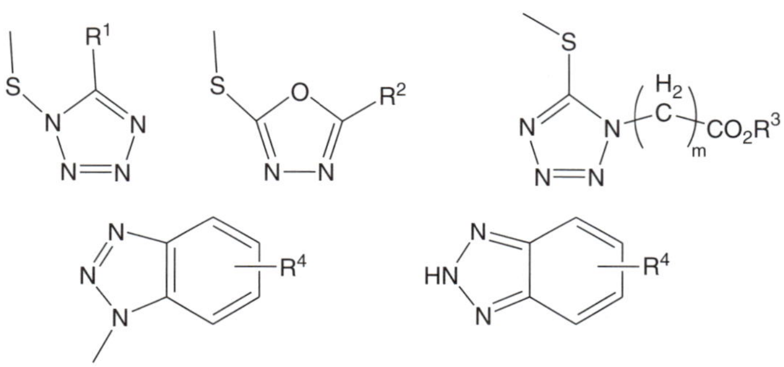

Figure 24 *Some examples of development inhibitor fragments*

inhibiting coupler must be similar in the rate at which it reacts with oxidised colour developer compared with the imaging coupler. In their 2006 patent, Singer *et al.*[5] describe some of the development inhibitor fragments used by Eastman Kodak Co., Figure 24.

The Fuji Photo Film Co. describes some of the yellow DIR couplers that they have patented,[6] Figure 25.

Figure 26 outlines the synthesis of one of the yellow couplers shown above in Figure 25.

DI(A)R couplers can contain many different features, depending on the application. All DI(A)R couplers however have a timing group which is attached at one end to the coupler and the other end to the inhibitor moiety.

The Eastman Kodak Co. products use a variety of timing groups,[7] see for example Figure 27.

Aromatic timing groups

In his patent mentioned above, Lau provides the synthesis of a cyan DI(A)R, Figure 28.

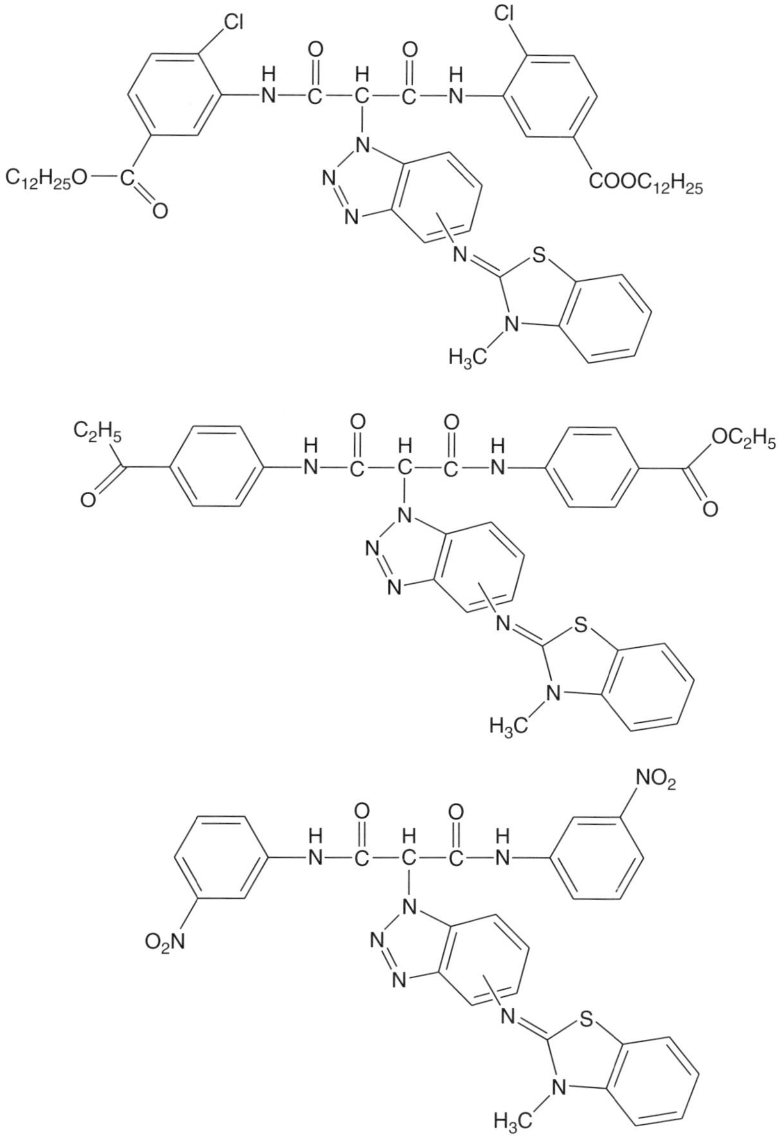

Figure 25 *Examples of Fuji Photo Film Co DIR couplers*

8.3 Oxidised Colour Developer Wandering

Colour fidelity can be compromised by a variety of effects that can take place between layers. Figure 29 shows the effect of oxidised colour developer migrating from one layer to another. In the left hand diagram there may or may not be problems with colour reproduction as the layers are sensitive to the same wavelength of light. In the right hand

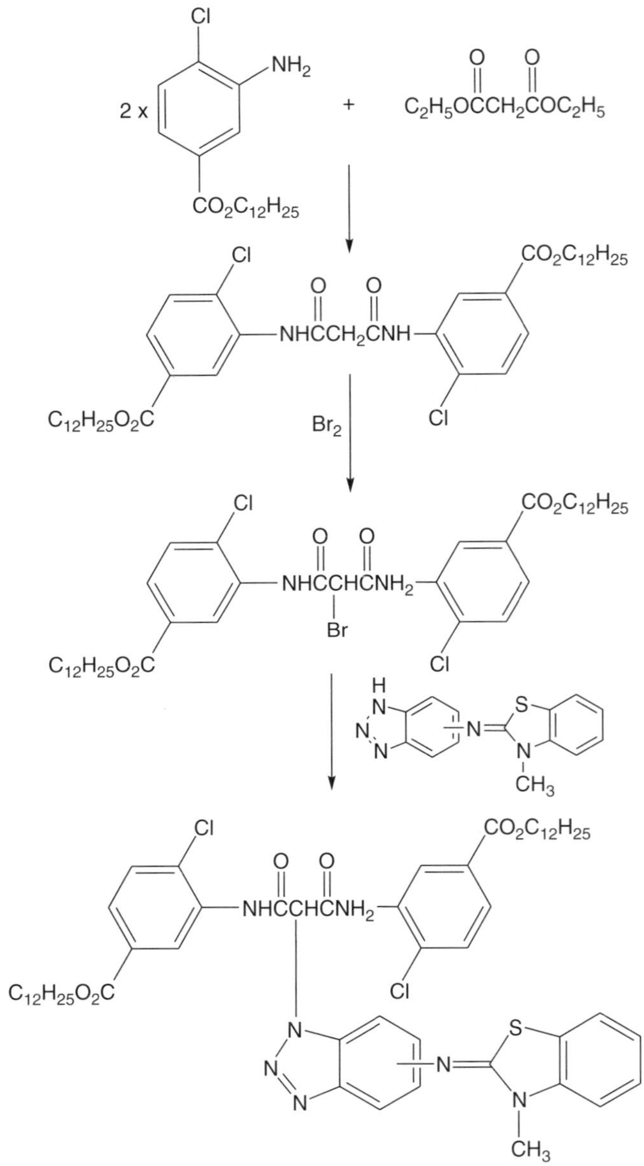

Figure 26 *The synthesis of a yellow DIR coupler*

diagram oxidised colour developer was generated in a cyan dye producing layer, followed by migration to a magenta dye producing layer.

Colour contamination will undoubtedly occur in the right hand diagram of Figure 29. The method employed to prevent the migration of oxidised developer between layers containing couplers is to coat an

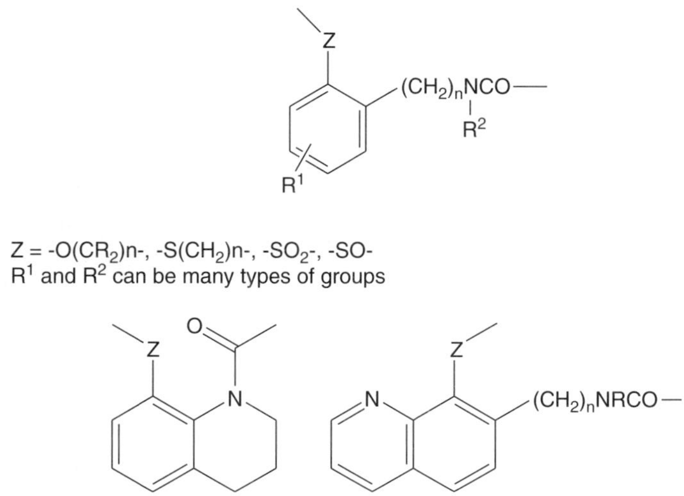

Figure 27 *Timing groups*

interlayer between the two colour records and incorporate an oxidised developer scavenger, Figure 30.

Oxidised colour developer scavengers are sometimes known as 'anti-stain' or indeed 'anti-colour mixing agents'. They work by reducing the oxidised colour developer that they come into contact with back to the developer form. They can be derivatives of amines, ascorbic acids, aminophenols, gallic acids, hydroquinones or hydrazines. Figure 31 shows two typical oxidised developer scavengers used by Eastman Kodak Co.

The use of polymeric hydroquinones as oxidised colour developer scavengers have also been patented, see for example ref 8, Figure 32.

8.4 Yellow Filter Layers

Chapter 3 discussed silver halide crystals and their natural sensitivity to blue light. Colour contamination of colour films will also occur if blue light is not filtered out, after exposure of the blue sensitive silver halide crystals, and before the green and the red sensitised silver halide crystals. This aspect of colour contamination is mentioned here for completeness and will be further discussed in Chapter 9, during the discussions of full film multilayer formulations. The diagram on the left in Figure 33 outlines the effect of light if no blue light protection is in place, and the right hand diagram the use of a yellow filter layer.

Figure 33 is a representation of the result of the absence and presence of a yellow filter layer. Clearly the layers would need to be exposed and processed in order to see the effects. Additionally, only two light sensitive

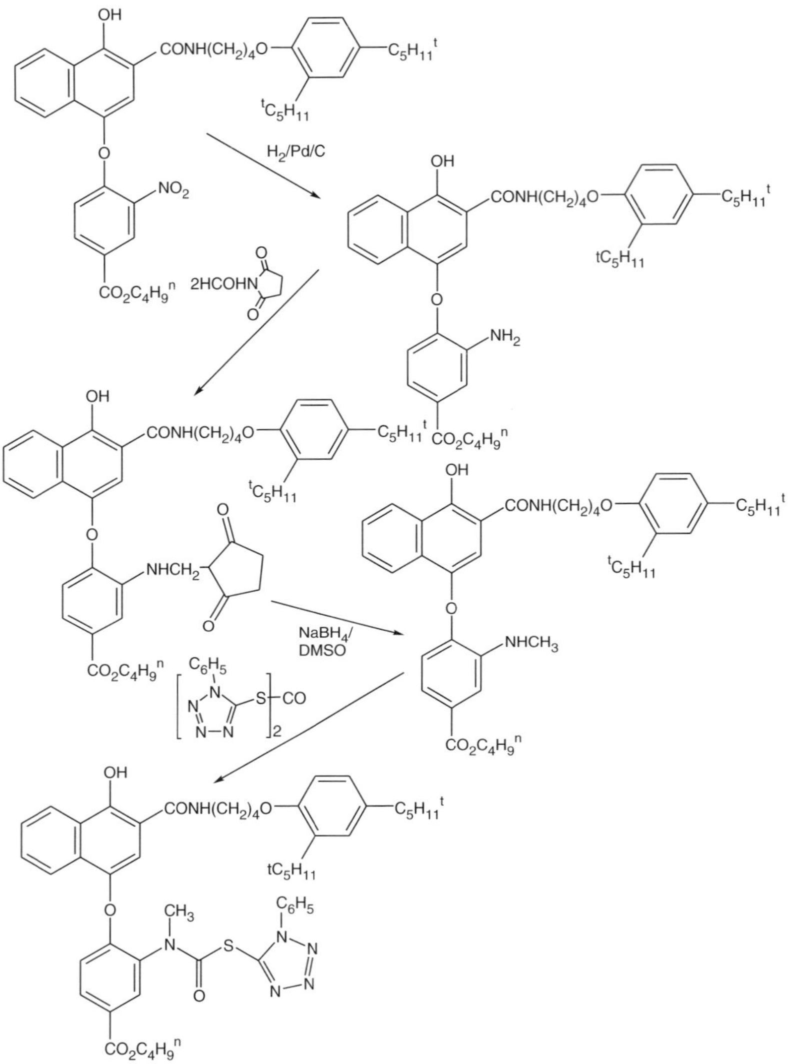

Figure 28 *An Eastman Kodak Co DI(A)R coupler synthesis*

layers are present. The red light sensitive layers would also be affected in the left hand diagram of Figure 33. Yellow filter layers need to be present in colour negative films during the exposure, but decolourise prior to printing. The obvious mechanisms therefore rely on the processing solutions to remove the yellow material.

Various materials have been patented for use in the yellow filter layer. A 1941 patent from Eastman Kodak Co.[9] suggested that a form of silver may be used

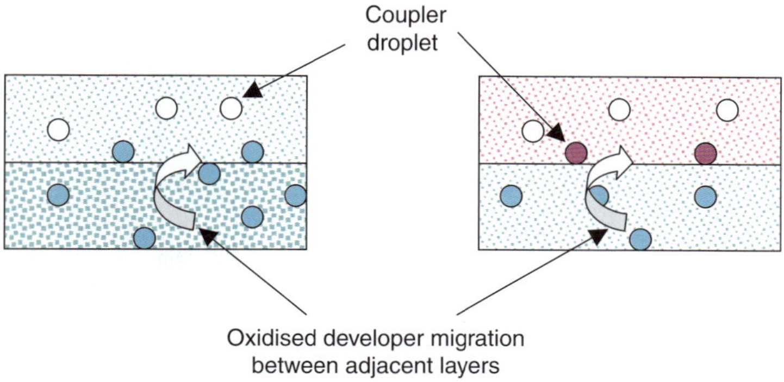

Figure 29 *Unwanted dye formation caused by D_{ox} wandering*

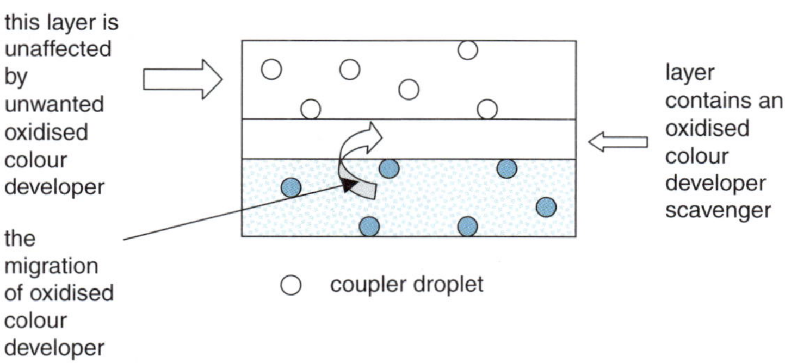

Figure 30 *Trapping D_{ox} in the interlayer using a D_{ox} scavenger*

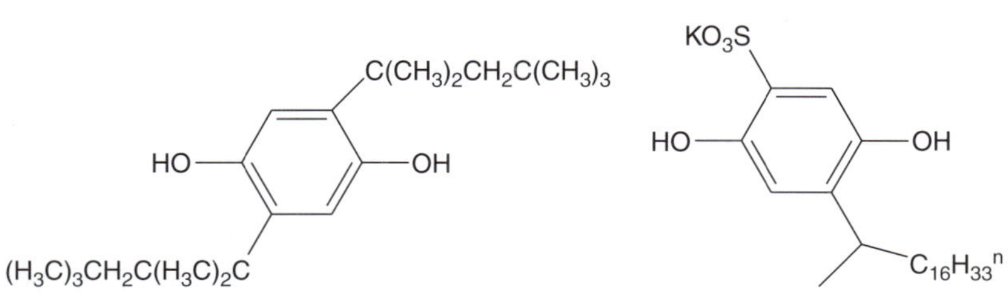

Figure 31 *Examples of D_{ox} scavengers*

. . . In silver dyestuff bleach-out processes it is known to provide intermediate layers, for example of plain gelatin, between adjacent emulsion layers for various purposes such as to carry filter colouring matters designed to protect under layers from the action of light which they are not destined to record. It has been proposed to incorporate in such intermediate layers, substances,

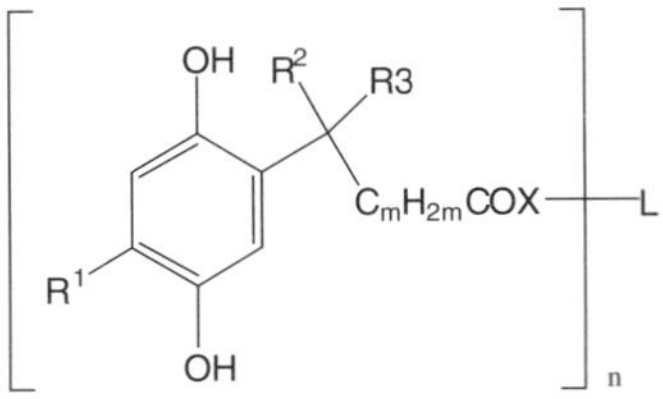

where n is always > 1, for example 2-100, preferably 2-50, especially 2-30 and in particular 2-4; X is O or NR^4;

If n=2

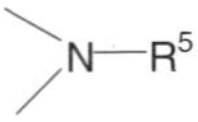

L is C_2 - C_{25}, alkylene; $C_4 - C_{25}$ – alkylene interrupted by O, S or C5 – C12 cycloalkylene or a phenylene or naphthalene

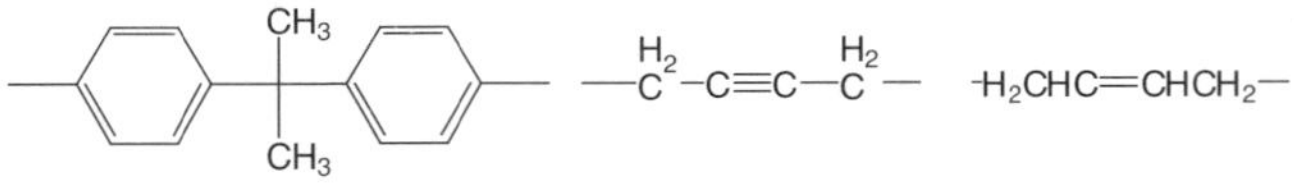

each of which may be substituted or unsubstituted by C_1-C_4 alkyl;

If n=3

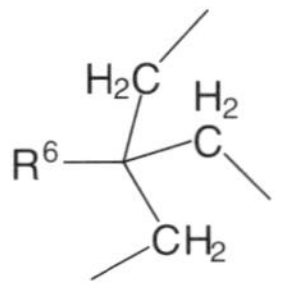

L is

If n = 4

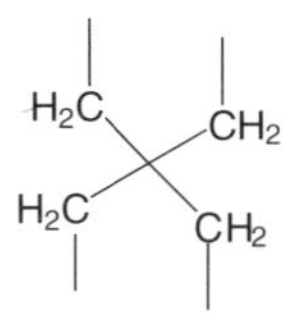

L is

Figure 32 *Examples of polymeric D_{ox} scavengers*

> *such as colloidal silver, which destroy or decolourise directly or in combination with the treating baths any dyestuff creeping into these layers from adjacent layers ... according to the present invention, therefore, in a method of colour processing multi-layer colour photographic elements by the silver dyestuff bleach-out process, we decolourize or destroy a filter dye and oxidize metallic silver both present together in an intermediate layer by means of the silver dyestuff bleach-out bath while it is effecting image-wise removal or destruction of dyestuffs in the emulsion ...*

The patent then suggests that suitable quantities of metallic silver are *'for the yellow filter 0.1 to 0.007 grams per square foot'*.

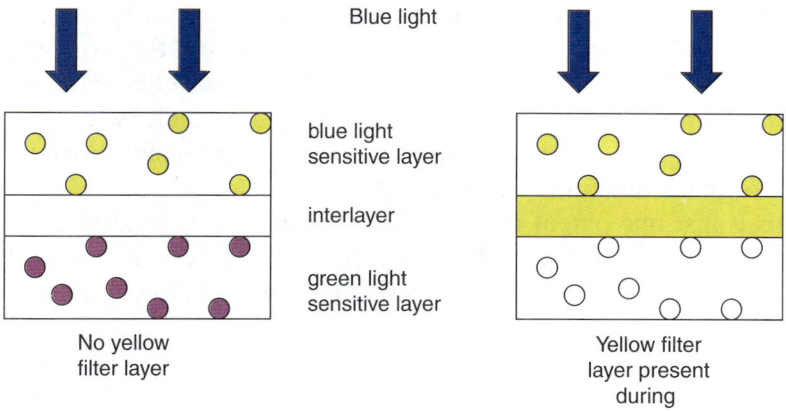

Figure 33 *The effect of yellow filter layers*

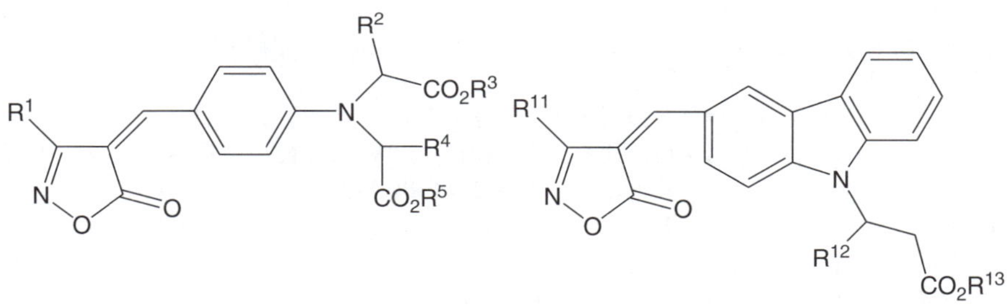

Figure 34 *Alternatives to colloidal silver for use in yellow filter layers*

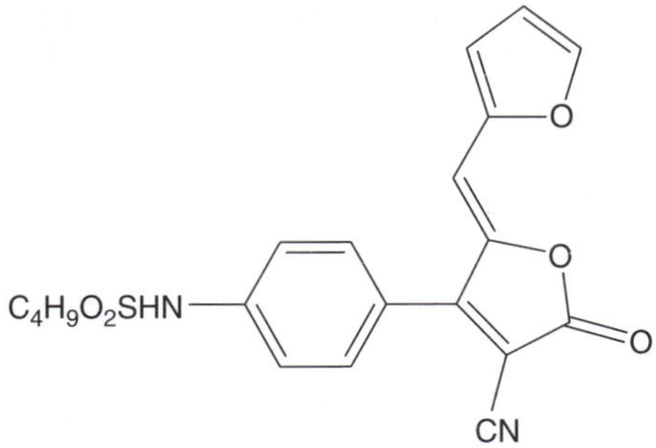

Figure 35 *An alternative filter dye from Eastman Kodak Co.*

There are some drawbacks to the use of colloidal silver (sometimes known as Carey-Lea silver), however, which include

- a ratio of green to blue density higher than desirable
- that colloidal silver can act as centres for physical development.

Agfa-Gavaert patented organic dye alternatives to the Carey-Lea colloidal silver, Figure 34, which they prepared by the condensation products of 3-alkylisoxazolones with *p-N,N*-biscarbalkoxymethylaminobenzaldehydes or *N*-carbalkoxyethylcarbazole-3-aldehydes.

Crawley *et al.*[10] report the use of an organic yellow filter dye in their study of masking couplers, Figure 35.

This is just one patent that has reported the use of an organic yellow filter dye as a substitute, which suggests that there is a general move away from colloidal (Carey-Lea) silver for use in yellow filter layers.

References

1. R.W.G. Hunt, *The Reproduction of Colour*, 4th edn, Fountain Press, Kingston-upon-Thames, 1987, ISBN 0-85242-356X.
2. H. Imamura, S. Sato, T. Kojima and T. Endo, US 4,070,191, Konishiroku Photo Industry Co. Ltd.
3. K. Shiba, T. Hirose, A. Arai, A. Okumura and Y. Yokota, US 4,163,670, Fuji Photo Film Co. Ltd.
4. R.L. Orvis, US 4,004,929, Eastman Kodak Co.
5. S.P. Singer, C. Grote, R.C. Stewart, J.W. Harder, D.S. Ross, J.N. Younathan and L.E. Friedrich, US 2006/0008751, Eastman Kodak Co.
6. M. Tanaka, M. Yagihara, T. Aono and T. Hirose, US 4,149,886, Fuji Photo Film Co. Ltd.
7. P.T.S.Lau, US 4,248,962, Eastman Kodak Co.
8. S. Jeganathan and S. Biry, WO 00/39064, Ciba Specialty Chemicals Holding Inc.
9. GB 540,969, Eastman Kodak Co.
10. J.N. Younathan, M.W. Crawley and K. Chari, US 6,132,943, Eastman Kodak Co.

CHAPTER 9

Film Structures

Film base is typically made of either polyethylene terephthalate or cellulose triacetate. Each base type can be made in a variety of thicknesses and widths depending upon the particular application. Polyethylene terephthalate bases are more often used for sheet films as they have good dimensional stability and are easier to handle in large formats. A typical method for the production of this base type is to cast the raw base in the following manner[1]

(a) *casting a molten polyethylene terephthalate resin having an inherent viscosity from 0.5 to 0.8 onto a casting surface,*
(b) *orienting by stretching in the machine direction at a stretch ratio between 2.0 and 4.0, and a temperature between 70°C and 130°C,*
(c) *tentering by stretching in the transverse direction, at a temperature between 70°C and about 130°C, and a stretch ratio between 2.0 and 4.0,*
(d) *heatsetting with constraint as in step (c) at a temperature between 200°C and 250°C, and*
(e) *detentering at the temperature of step (d) by permitting the width to shrink from 2 to 20% to achieve a film having a planar birefringence from 0.12 to 0.149.*

Having achieved the required width and thickness, various layers are coated onto one or both sides of the base. Required physical properties of the fully finished base include

- dimensional stability
- mechanical strength
- resistance to curling
- resistance to thermodeformation

- resistance to water
- transparency.

The finishing operation, which is undertaken after the base has been coated with photographic layers, also requires that the base is suitable for

- chopping
- cutting
- perforating.

The nature of the base undercoat varies, depending upon the type of film, which is to be manufactured. These layers can be coated on the raw base using Gravure coating methods – see Figure 1.

A typical example of the type of base undercoat that has been coated (at least in an experimental film, for a motion picture film application) can be found in the patent mentioned above.[1] The undercoat described in this patent has been designed to facilitate adhesion of the emulsion layers to the raw film base. The undercoat consists of a terpolymer of acrylonitrile, vinylidene chloride and acrylic acid – see ref 2 for the method of preparation. This film was coated on the backside of the film, with a carbon black dispersion in a cellulose acetate naphthalene binder. The emulsion layers were coated on the front side.

In another patent, Greener and Wen-Li Chen[3] evaluated the use of mixtures of poly(ethylenenaphthalate) and poly(ether imide). They describe the use of 70–95 weight per cent of the former monomer and 30–50

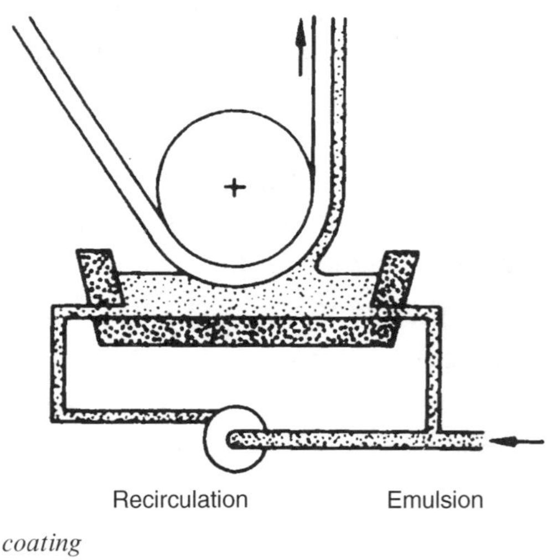

Figure 1 *Gravure coating*

weight per cent of the latter. This particular base formula has improved core-set curl, low post-process curl, good stiffness and good tear strength.

Colour negative films use cellulose triacetate film base. This base type is often made by solution casting where dopants are added to tint the base. A schematic for the production of this base appears below as Figure 2.[4]

The use of cellulose triacetate as a film base is not restricted to the Eastman Kodak Co. (for example see ref 5) however, as other manufacturers have also used those types of polymers, see for example.[6]

Film coating involves the use of multi-slide hoppers that deliver the relevant solutions onto the pre-prepared base. The figure below shows a simplified cross-section of bead coating, Figure 3.

Figure 3 shows a film coating with an anti-halation undercoat layer (depicted in grey) coated on the base at the same time as the cyan, magenta and yellow layers. In practice there would be interlayers between the cyan and magenta, and between the magenta and yellow

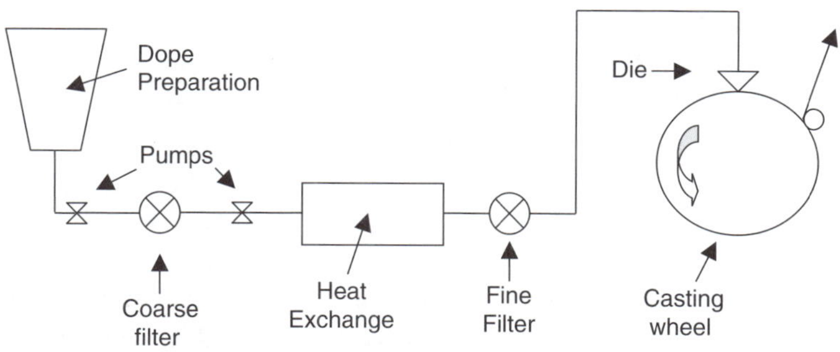

Figure 2 *Solution casting of base*

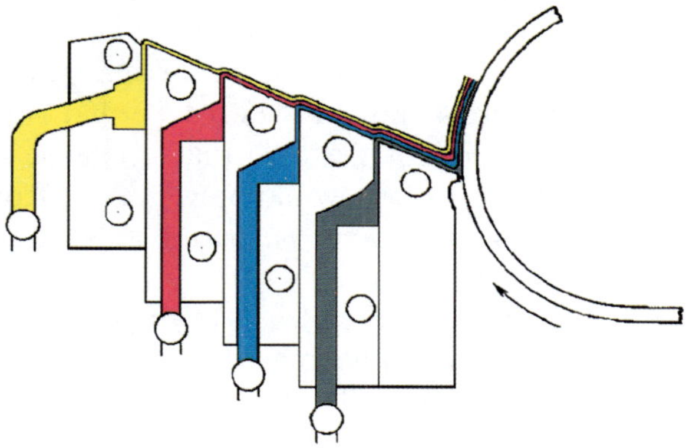

Figure 3 *Multi-slide hopper coating*

layer so that there is no chemical cross-contamination. In addition there would be either a combined supercoat and UV protection layer or two layers, one for UV protection and one for abrasion protection, coated above the yellow layer.

There are several points to note

- Several layers are coated at the same time requiring a viscosity profile between the layers with the bottom layer being of lowest viscosity.
- The addition of a surfactant as a coating aid to the uppermost layer. This layer is the first to make contact with the base when the coating is started and so this layer needs good wetting properties. In addition the layers are stretched as they leave the hopper slide and travel up the web.
- This process is undertaken in darkroom conditions as the photographic product would be exposed should there be any light.
- A vacuum is needed at the point between the hopper and coating roller to maintain a good and stable coating 'bead'.
- As the solutions travel down the hopper, there is a need for a physical barrier to help maintain the edge profile.

9.1 Coating Aids

Many surfactants have been evaluated as coating aids, some cited in the patent literature. Surfactants used in this way are designed and used in the coating of uniform layers with a minimum of coating defects. A repellency is a coating unevenness such as an oval, round or comet shaped indentation or crater in the layer, or layers, being coated. These defects are usually caused either by the presence of insoluble material in the coating solutions themselves, or contaminants on the surface to be coated. Surfactants also help in producing coatings that are free of defects at the edges. Figure 4 shows a defect type that is influenced by the use of surfactants. It is not the case that a change to either the surfactant laydown or composition would prevent these from occurring, more that the removal of surfactants as coating aids would render the solutions incapable of being coated.

A large number of compounds are used as coating aids throughout the various products of all photographic manufacturers. It is not the intent here to create a list of commercially available coating aids. There are some compounds that have been prepared and evaluated which are unique to each manufacturer. Pitt, Clark and Padday[7] described the use

Figure 4 *A coating defect*

$CF_3(CF_2)_nCH_2O(glycidyl)_mH$

Where n is an integer from 4 to 7 and m is an integer from 6 to 45.

Figure 5 *Fluoro-surfactants used as coating aids*

of fluoro-surfactants, Figure 5. This material and other fluoro-surfactants are bespoke and therefore not available commercially.

9.2 Film Structures

Three film types have been chosen from the vast number that exist, as examples of the types and number of layers that are used in the final products. Arguably, a colour negative film is the most complex photographic product, and so a typical example from the patent literature will be examined in detail, so that all the compounds that have been described in previous chapters can be placed in context. Perhaps the most simple product in terms of the number of layers and chemicals, is a black and white film, the structure for which is provided as shown in Figure 6.

Black and white films typically have three layers, two of which contain emulsions. The two emulsion layers contain an emulsion with a different mean grain size distribution. Figure 7 represents one of the emulsions with a spread of silver halide grain sizes.

The larger grains on the right of the distribution curve will have a higher photographic speed. This photographic speed manifests itself on the density *vs.* log exposure plot as outlined in Figure 8.

The grains near 'A' in Figure 8 are the largest grains in the size distribution and the point 'B' are the smallest. A larger photographic exposure range, or latitude, can therefore be achieved using emulsions that have a different size distribution. Mixing two emulsions of different size distributions into the same melt, and therefore the same layer, will

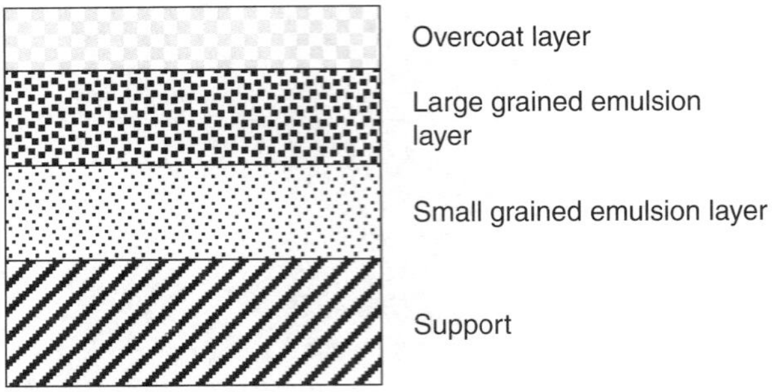

Figure 6 *The layer order of a typical black and white film*

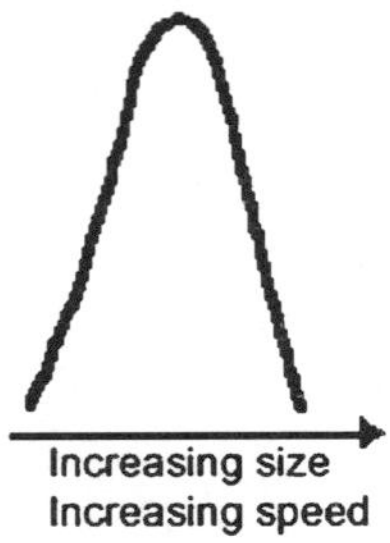

Figure 7 *Silver halide grain size distribution*

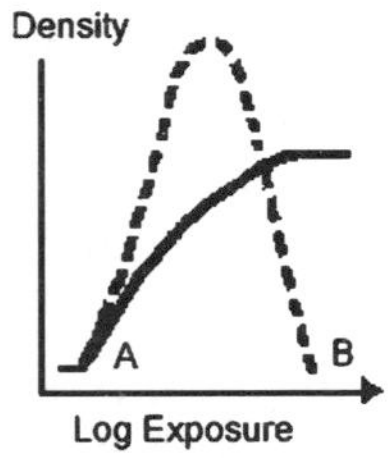

Figure 8 *The correlation of silver halide grain size and photographic speed*

achieve the photographic latitude, but will remove several degrees of freedom as the average granularity and sharpness for example will also be compromised. The two layers are therefore contributing to two different parts of the density log exposure curve, see Figure 9.

The contribution of both the fast, larger grained emulsion, and the slower, smaller grained emulsion, to the overall black and white density *vs.* log exposure curve has been shown in Figure 9. The fast layer is coated nearer the film surface so that it is closer to the light source, which will expose the larger grains first. It is therefore the larger grains that contribute to the photographic speed, which is measured just above the inflection point, or toe, Figure 10.

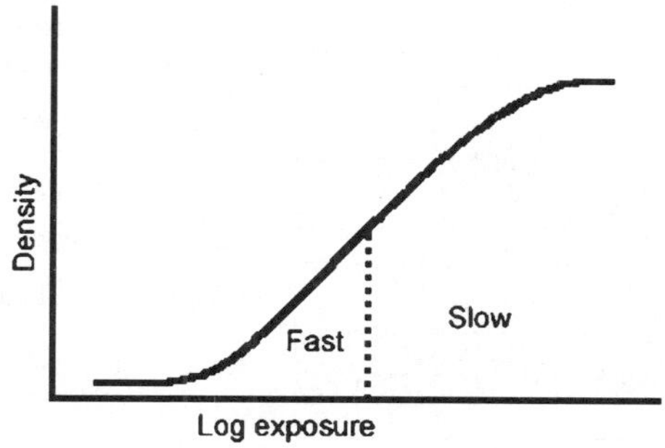

Figure 9 *The contribution of the two layers to the overall curve shape*

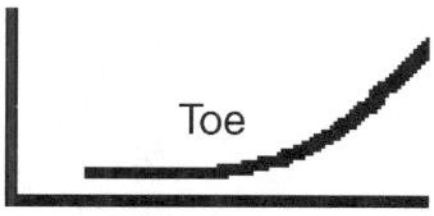

Figure 10 *The area of the density vs log exposure curve known as the toe*

The issue at hand is precipitating two emulsions that have a desired speed separation. This allows for a smooth transition from one emulsion to the other on the straight-line portion of the density *vs.* log exposure plot.

A colour negative film, at least up to and including the early 1970s, was to some extent similar to the sum of three black and white film packs as outlined above in Figure 6. Two layers were used for each colour record, with interlayers between the red and green sensitive layers, and a yellow filter layer to prevent the issues discussed in Chapter 8, Figure 33. In later years some film manufacturers increased the number of layers per film record to three. Perhaps the first issue is why are there so many layers in a colour negative film? Figure 11 shows a schematic diagram of a typical coating structure.

9.3 Anti-Halation Undercoat (AHU) Layer

All photographic products suffer from internal reflections and refractions once light has entered the uppermost layer. The light that is reflected or refracted is but a small percentage of the available light entering the system. A problem that causes the greatest concern is that of light reflected from the gel/base interface at the bottom of the coated layers, Figure 12.

If left unchecked, some of the incident light which reaches the film or paper base will be reflected back into the multilayer. This could cause

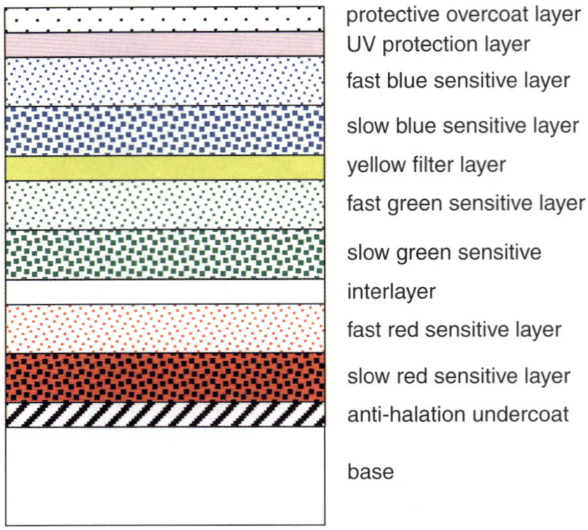

Figure 11 *A typical colour film multilayer*

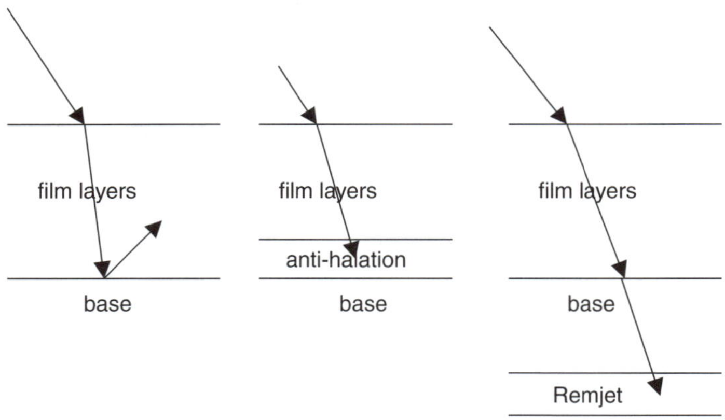

Figure 12 *Trapping unwanted light*

unwanted exposures at a site that might be some distance from the point of the original exposure. This will definitely affect sharpness, and may affect granularity, as the dye cloud thus formed may be isolated from the image dye clouds. This could cause a dye density issue in both the image area and the area of the actual exposure. A method of controlling this back reflected light is to coat an additional layer, the purpose of which is to absorb this light. This is called an anti-halation layer and in the case of film products can be coated on the backside of the film and removed during processing (right hand diagram in Figure 12), or coated as the first layer of the film product, as in the middle diagram.

More details of the right hand case appear in Chapter 12 during the discussion of motion picture films. In the case above, the film structure in Figure 11 does not allow for physical removal as the anti-halation layer is integral to the film 'pack'. In this case a chemical that is present during exposure, but is removed by the processing solutions, is the only option. In many cases, the material used for anti-halation protection in the AHU layer is a form of black filamentary silver, which is precipitated just for this purpose.

The slow and fast red sensitive layers follow the pattern described above for the black and white films. In this case there is an added issue compared with black and white films, as there is the need to be mindful of the ratios of the various chemicals as well as the content, see Figure 13.

Figure 13 shows only silver, coupler droplets and the eventual dye clouds. None of the competing couplers such as DIRs or coloured couplers are shown. Under normal circumstances, all the silver will have been removed during processing. They have been left in the diagram to provide an orientation for the exposure point.

The left hand diagram shows many more dye clouds per developed silver grain, compared with the right hand diagram. In both cases the amount of oxidised colour developer generated from the reaction between the latent image and developer is the same, merely that in the left hand case the oxidised colour developer has ready access to coupler molecules with which to react and form dye molecules. The left hand scenario will therefore produce a different granularity of the processed material compared with the right hand scenario.

The ratio of chemicals within the layer is therefore important and is another factor in the design of a multilayer – whatever the product is.

The next layer in Figure 11 is an interlayer, which will contain an oxidised developer scavenger. The two green sensitive layers are as

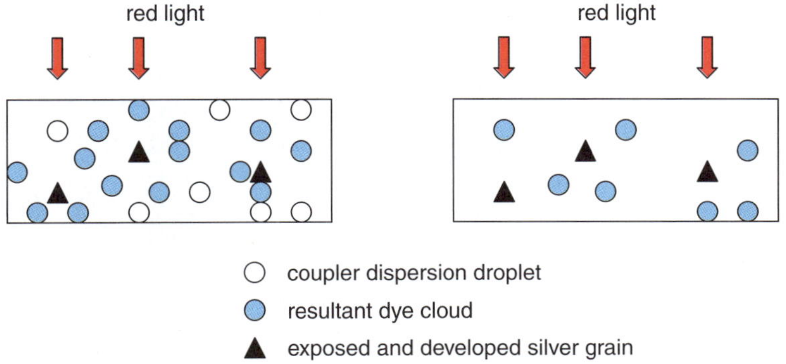

Figure 13 *The effect of differing levels of coupler laydown relative to that of silver halide*

discussed for the red sensitive layers. The yellow filter layer and its composition have already been discussed – see Chapter 8 Figure 33. The two blue sensitive layers are similar in composition to the green and red sensitive layers. The next layer in the structure is an UV protection layer.

9.3.1 UV Protection Layer

UV radiation causes harmful effects on the image dyes. Molecules capable of absorbing UV light are therefore added to the coating 'pack' to prevent unwanted exposures. Many different chemicals have been evaluated as UV protection compounds. Fuji Film Co. Ltd. have patented the use of materials such as those described in Figure 14.[8]

Eastman Kodak Co. have provided examples of different classes of compounds for UV protection including those described in Figure 15.[9]

The final layer outlined in Figure 11 is a protective overcoat layer.

9.3.2 Protective Overcoat Layer

There are many physical defects that can perturb the surface of a film or paper. Liquid damage is difficult to prevent, but there are issues that relate to abrasion sensitivity and static that can be addressed. An

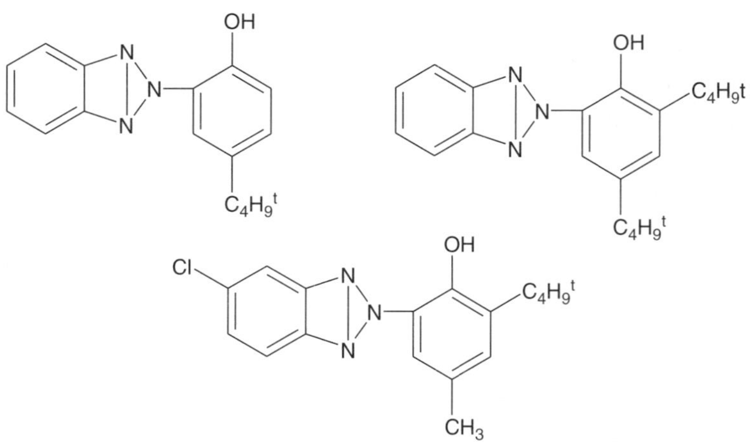

Figure 14 *UV absorbing compounds used by Fuji Photo Film Co Ltd*

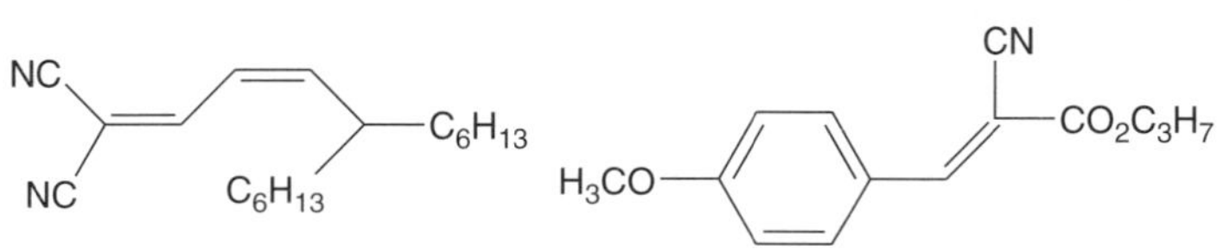

Figure 15 *Alternative UV absorbing compounds used by Eastman Kodak Co.*

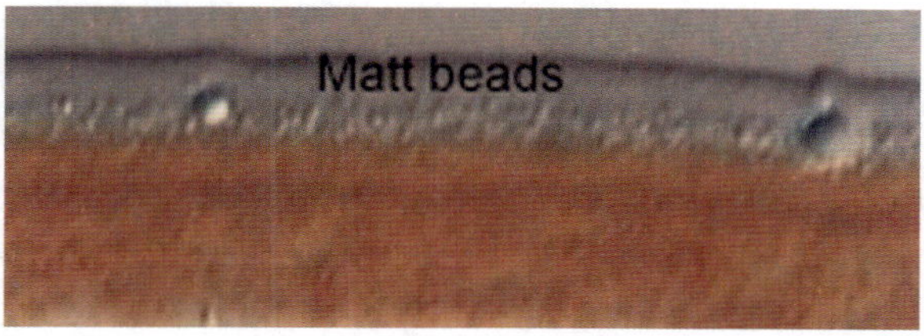

Figure 16 *A cross-section of the upper layers of a colour negative film after processing*

intimate contact between two surfaces can lead to peel apart static when those two surfaces are separated. In the case of unexposed film or paper, this can lead to unwanted exposure of the silver halide grains. Additionally some graphic arts films are used in one of the stages of printed circuit boards where they need to ensure intimate contact between surfaces during exposure of the film. This requires a good drawdown vacuum during which there needs to be a method of removing all of the air between the exposure platten and film surfaces. For all these reasons and others, polymeric beads are added to the supercoat or uppermost layer of a photographic product. Figure 16 shows a cross-section of an actual film product.

Three matte beads can be seen in this small cross-section of exposed and processed colour negative film (two are in intimate contact on the right hand side of the diagram). All photographic manufacturers employ this technology. Fuij Film B.V. have reported the use of acryl-modified copolymers of polyvinyl alcohol,[10] whereas Eastman Kodak Co. have described the use of polymethylmethacylate.[9] Fuji Film Co. Ltd. also comment on the preparation of matte particles designed for better adhesion during processing.[11,12] In a further patent Eastman Kodak Co. report matte beads which have a polymeric core surrounded by a layer of colloidal inorganic materials.[13]

There is no advantage of coating matte beads in any layer other than the uppermost layer. On first inspection, therefore, it seems somewhat unusual that some of the colour films produced by the Fuji Film group appears to have matte beads part way through their film packs. The inference is that the matte beads are added to the top layer of a coated pack, and that this pack had some further layers coated over the top of the beads. This would imply a multi-pass coating operation.

9.4 Colour Film Latitude

'Film latitude' is the term used to describe the property of a film, which allows for the capture of an image should a photographer decide to under or over expose the film negative.

The required colour film latitude, at least for the amateur film market, is such that two emulsions, one coated in each of the two layers is insufficient. These types of films often have three emulsions and sometimes four. In the above schematic three emulsions were used for each of the three colour records, but the schematic, Figure 11, only shows two layers per colour record, one emulsion in each of the fast layers, but two emulsions in each of the slow layers. Figure 17 shows this use of emulsions for the red record.

The fast layers usually contain one emulsion that is of the appropriate size and spectral sensitisation (see Chapter 3) for the speed of the film. The slow layer usually contains a mixture of a medium sized emulsion grains and a slow emulsion grains that differ in their speed by virtue of the mean grain size, sometimes known as equivalent circular diameter (ECD), of the silver halide crystals.

Figure 18 is a cross-section of an actual coated unexposed colour negative film.

Figure 19 shows the fast layers that have been exposed under reduce lighting conditions. Note that the yellow filter layer and anti-halation undercoat layer are now optically inert. This was achieved during the processing step.

Figure 20 shows a cross-section of the same film that has been fully exposed and processed.

Colour negative film design is always a matter of compromise. An improvement in granularity (measurement of perceived graininess) can be achieved by increasing coupler weight by changing the coupler to one which will produce a higher extinction coefficient dye with oxidised colour developer and by changing the degree of coupler starvation – see above

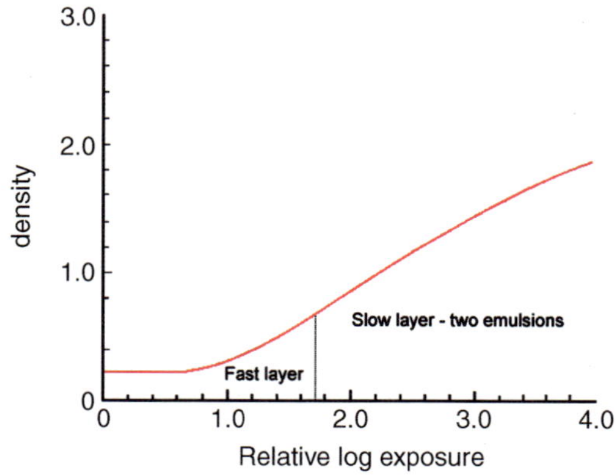

Figure 17 *The contribution of three silver halide emulsions coated in two layers to the eventual curve shape*

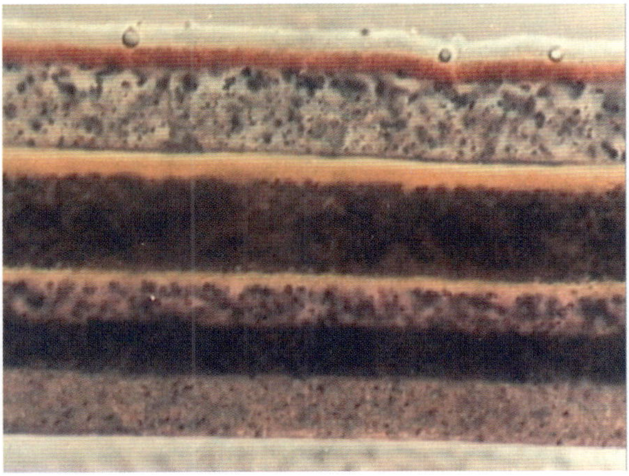

Figure 18 *A cross-section of unprocessed colour negative film*

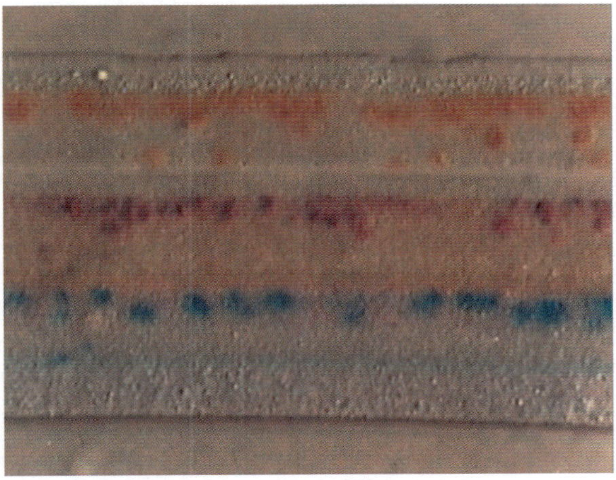

Figure 19 *Partial exposure of the negative*

Figure 13, *etc*. One of the major factors is silver halide laydown. An increase in the emulsion weight per unit volume would make the film more expensive as there would be a greater number of moles of silver used in the film. While an emulsion weight per unit area would improve the granularity of the layer in question, there would be a penalty to the underlying layer, as there would be more light scatter caused by the increase in the number of light scattering centres (silver halide grains), Figure 21.

This 'triangle' expresses the frustration that you can always influence the design of the film for two of the three components in a positive way, however the third point of the triangle will always be adversely affected.

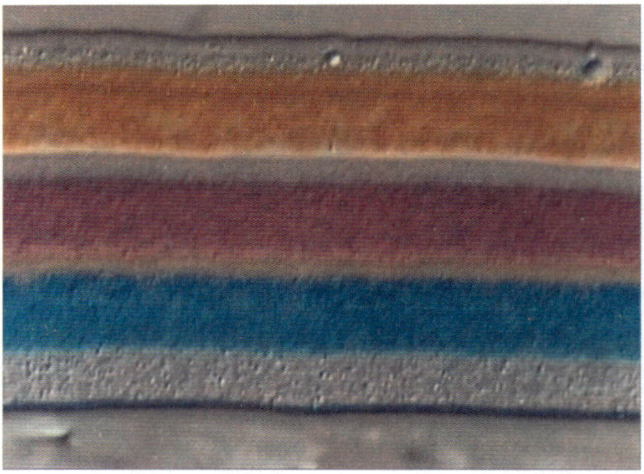

Figure 20 *Fully exposed and processed negative*

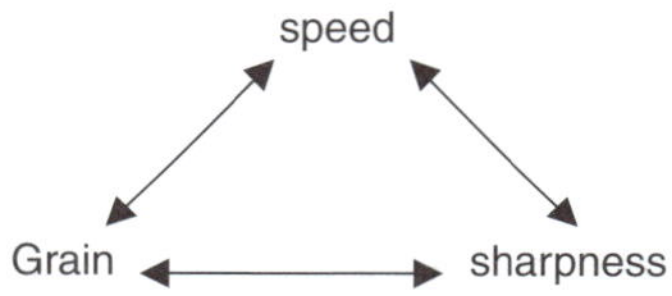

Figure 21 *The speed/grain/sharpness triangle*

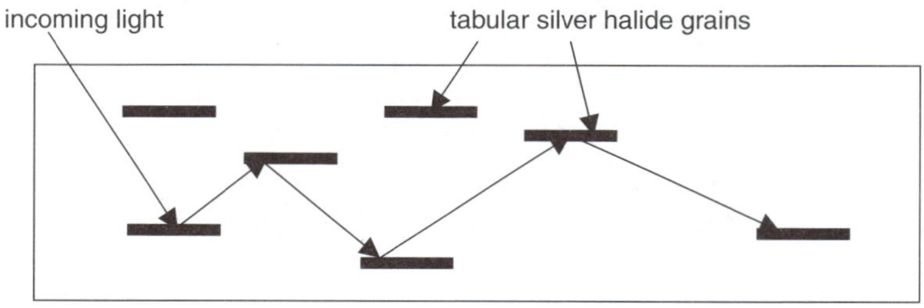

Figure 22 *Potential 'light-piping' caused by silver halide grain alignment*

Two technologies over the past 30 years or so have, so to speak, increased the area of the triangle. In their own way, each technology has provided a means of improving the granularity or speed of a film, without a detrimental loss in sharpness. The first of these technologies to be described here, although the second chronologically, was the introduction of tabular emulsion grains – see Chapter 3 for further details.

The issue of light scatter with thin tabular silver halide grains, which was alluded to above, is related to the potential alignment of these thin grains within a layer, leading to 'light piping' under some circumstances. In Figure 22, one layer of a multi-layer coating has been shown so that the phenomenon can be exaggerated in order to demonstrate the principle. Despite this relatively minor issue, the implementation of tabular emulsions has been seen in most of the colour negative films for most of the manufacturers.

The second of the two methods of creating a different 'photo-space' *i.e.* to a certain degree breaking out of the conventional speed/grain/sharpness triangle was pioneered in the late 1960s and early 1970s by Eeles and O'Neill[14] of Kodak Limited. Their premise was that there was no reason to keep to the film structure outlined in Figure 11, and pioneered the use of a so-called inverted structure – Figure 23. In this film structure the fast red layer has been placed between the fast green layer and the slow green layer. Additional interlayers have been introduced to prevent the effects of oxidised colour developer wandering between the red and green layers.

All other things being equal, the fast red sensitive layer will now be exposed to more of the available light at the same exposure, as the layer is closer to the top of the film pack. The film builder now has choices. He/she can

- reduce the fast red sensitive emulsion grain size which will reduce the speed of the layer and keep the silver halide coverage constant or
- keep the fast red layer silver halide laydown the same and add a dye to reduce the speed of the layer.

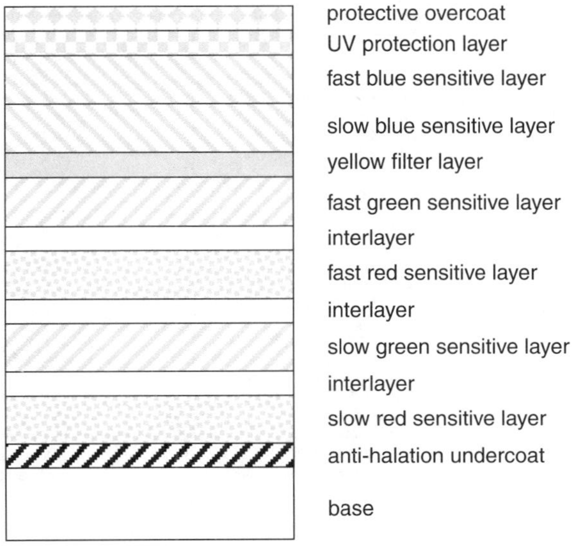

Figure 23 *An example of an 'inverted' layer order*

Eeles and O'Neill provide some examples in their patent that illustrates this concept, some data from which is reproduced below. Figure 24 shows the laydown rates for the check coating, which was coated as a control coating with the conventional coating format, and Figure 25 shows details of two of the experimental films, described as film 4 and 5 in the patent.

Film 3 is a control film, which was coated to the conventional film structure and was used to assess the impact of layer order changes.

The flowing couplers were used in these coatings with references to further details for each of the relevant couplers.

Coupler 1 – Cyan dye-forming coupler[15,16]

- 1-Hydroxy-2-[δ-(2,4-di-tert-amylphenoxy)-*n*-butyl]naphthamide
- 1-Hydroxy-2-[β-(2,4-di-tert-amylphenoxy)ethyl]naphthamide
- 5-[α-(2,4-di-tert-amylphenoxy)hexanamido]-2-heptafluoro-butyr-amidophenol
- 1-Hydroxy-2-[β-(2,4-di-tert-amylphenoxy)-*n*-butyl]naphthamide.

Layer	*Component*	*Check coating – called film 3 in the patent*
Protective gelatin overcoat	*Common layer for all experiments*	*Common layer for all experiments*
Blue sensitive, yellow dye forming layer(s)	*Common layer for all experiments*	*Common layer for all experiments*
Yellow filter layer	*Common layer for all experiments*	*Common layer for all experiments*
Fast green sensitive magenta dye-forming layer	*green sensitive AgBrI magenta dye-forming coupler 4 yellow coloured, magenta-forming gelatin*	*16 mg/dm^2 1.3 mg/dm^2 0.62 mg/dm^2 21.5 mg/dm^2*
Slow green sensitive magenta dye-forming layer	*medium speed, green sensitive, AgBrI slow green sensitive AgBrI magenta dye-forming coupler 4 yellow coloured, magenta dye-forming coupler 7 gelatin*	*4.8 mg/dm^2 9.1 mg/dm^2 4.3 mg/dm^2 2.1 mg/dm^2 21.5 mg/dm^2*
Interlayer	*Gelatin*	*8.9 mg/dm^2*
Fast red sensitive cyan dye-forming layer	*fast red sensitive AgBrI cyan dye-forming coupler 1 magenta coloured, cyan dye-forming coupler 2 gelatin*	*18.8 mg/dm^2 1.6 mg/dm^2 0.18 mg/dm^2 21.5 mg/dm^2*
Slow red sensitive cyan dye-forming layer	*medium speed, red sensitive AgBrI slow red sensitive AgBrI slower red sensitive AgBrI cyan dye-forming coupler 1 magenta coloured, dye-forming gelatin*	*8.6 mg/dm^2 4.3 mg/dm^2 4.3 mg/dm^2 6.5 mg/dm^2 0.72 mg/dm^2 221.5 mg/dm^2*
Film Support	*Film Support*	*Film Support*

Figure 24 *Laydown rates for the internal control coatings*

Coupler 2 – Cyan dye-forming coloured coupler[17]

- 1-Hydroxy-4-phenylazo-2-[4′-(*p*-tert-butylphenoxy)]naphthamide
- 1-Hydroxy-4-(4-[2-{8-acetamido-1-hydroxy-3,6-disulfo-naphthyl}azo]phenoxy)-2-(α-[2,4-di-tert-amylphenoxy]butyl)naphthamide
- 4-(2-Acetylphenylazo)-1-hydroxy-2-[δ-(2,4-di-tert-amylphenoxy)-*n*-butyl]naphthamide
- 1-Hydroxy-4-phenylazo-*N*-isoamyl-2-naphthanilide
- 4-(4-{7-[1-Acetamido-3,6-disulfo-8-hydroxylnaphthylazo]-phenoxy}-1-hydroxy-2-[δ-(2,4-di-tert-amylphenoxy)-*n*-butyl]naphthamide, disodium salt

Layer	*Component*	*Laydown of film 4*	*Laydown of film 5*
Protective gelatin overcoat	*Common layer for all experiments*	*Common layer for all experiments*	*Common layer for all experiments*
Blue sensitive, yellow dye forming layer(s)	*Common layer for all experiments*	*Common layer for all experiments*	*Common layer for all experiments*
Yellow filter layer	*Common layer for all experiments*	*Common layer for all experiments*	*Common layer for all experiments*
Fast green sensitive magenta dye-forming layer	*green sensitive AgBrI magenta dye-forming coupler 4 yellow coloured, magenta dye-forming coupler 7 gelatin*	*16 mg/dm^2 2.6 mg/dm^2 1.25 mg/dm^2 21.5 mg/dm^2*	*16.1 mg/dm^2 2.6 mg/dm^2 1.2 mg/dm^2 21.5 mg/dm^2*
Interlayer	*Gelatin*	*8.9 mg/dm^2*	*8.9 mg/dm^2*
Fast red sensitive cyan dye-forming layer	*fast red sensitive AgBrI cyan dye-forming coupler 1 magenta coloured, cyan dye-forming coupler 2 gelatin*	*21.5 mg/dm^2 3.2 mg/dm^2 0.35 mg/dm^2 21.5 mg/dm^2*	*21.5 mg/dm^2 4.8 mg/dm^2 21.5 mg/dm^2*
Interlayer	*Gelatin*	*8.9 mg/dm^2*	*8.9 mg/dm^2*
Slow green sensitive magenta dye-forming layer	*fast green sensitive AgBrI medium speed, green sensitive, AgBrI slow green sensitive AgBrI magenta dye-forming coupler 4 yellow coloured, magenta dye-forming coupler 5 gelatin*	*3.2 mg/dm^2 7.5 mg/dm^2 8.6 mg/dm^2 6.6 mg/dm^2 1.1 mg/dm^2 21.5 mg/dm^2*	*7.5 mg/dm^2 11.8 mg/dm^2 6.5 mg/dm^2 1.1 mg/dm^2 24.2 mg/dm^2*
Interlayer	*Gelatin*	*8.9 mg/dm^2*	*8.9 mg/dm^2*
Slow red sensitive cyan dye-forming layer	*medium speed, red sensitive AgBrI slow red sensitive AgBrI slower red sensitive AgBrI cyan dye-forming coupler 1 magenta coloured, dye-forming coupler 2 gelatin*	*7.5 mg/dm^2 3.8 mg/dm^2 4.3 mg/dm^2 5.9 mg/dm^2 0.66 mg/dm^2 21.5 mg/dm^2*	*As film 3*
Film Support	*Film Support*	*Film Support*	*Film Support*

Figure 25 *Laydown rates for the experimental 'inverted layer' coatings*

Coupler 4 – Magenta dye-forming coupler[18,19]

- 1-(2,4,6-Trichlorophenyl)-3-[β-(2,4-di-tert-amylphenoxy- propion-amido]-5-pyrazolone
- 1-{4-[α-(3-t-butyl-4-hydroxyphenoxy)tetradecanamido]-2,6-di-chlorophenyl}-3-(2,4- dichloroanilino)-5-pyrazolone
- 1-(2,4,6-Trichlorophenyl)-3-[3-α-(2,4-di-tert-amylphenoxy)-acet-amido-benzamidol-5- pyrazolone
- 1-{4-[α-(3-t-butyl-4-hydroxyphenoxy)hexanamido]phenyl}-3-pen-tadecyl-4-carboxyphenoxy-5-pyrazolone
- 1-(2,4,6-Trichlorophenyl)-3-{3-[α-(3-t-butyl-4-hydroxyphenoxy)-tetradecanamido]benzamido}-4-phenylthio-5-pyrazolone

Coupler 7 – Magenta dye-forming coloured coupler[20]

- 1-(2,4,6-Trichlorophenyl)-3-{4-[α-(2,4-di-tert-amylphenoxy)-but-yramido]aniline}-4-(2-ethoxyphenyl)azo-5-pyrazolone
- 1-(2,4,6-Trichlorophenyl)-3-[3-(2,4-diamylphenoxy-acet-amido)benzamido]-4-(p-methoxyphenylazo)-5-pyrazolone

A comparison of the relative log speeds (0.1 above minimum density) for the three films was also given in the patent, Figure 26.

The blue speeds should be constant as they were common layers across the experiment, and show the variability within the experiment. Both red and green speed of film 4 showed a speed gain relative to the control. The principle difference between film 5 and film 4 was the omission of the magenta coloured cyan dye forming coupler from the fast red sensitive layer. In all cases, coupler and silver laydown levels needed to be adjusted to maintain the necessary final density *vs.* log E exposure curve.

There are other chemicals that were added to the multilayer films shown in Figures 24 and 25. For example the coating aids (surfactants) have been omitted. DIR and DI(A)R couplers were coated as couplers 2 and 7. Additionally the anti-halation undercoat layer is not discussed in the above example, as this layer was constant in each of the three experimental coatings.

Film	Red	Green	Blue
3	2.95	3.18	3.64
4	3.20	3.21	3.68
5	3.21	3.30	3.69

Figure 26 *A comparison of film speeds from the 'inverted layer' experiments*

The above example is the seminal patent concerning film layer order. This example serves to illustrate the complexity of the overall photographic system. A modern colour negative film may have 12–14 layers and over 100 chemicals.

While this film structure afforded extra options for the film builder, there is a trade-off in the manufacturing process as it may not be capable of coating the extra layers, at least in one pass through the coating machine. Some coating facilities have or had three coating stations each capable of delivering the relevant solutions to the coating point, then drying the layers prior to the next application. The greater the number of layers, the less likely it is that the manufacturing process will be able to coat all of the layers in one pass. Introducing multiple passes generates more waste during the manufacturing process and entails more waste on the ends of the production rolls and at coating starts and stops. Additionally, the final film sensitometry is more difficult to control if one colour record has been coated prior to the other layers. In this case no adjustments can subsequently be made to the pre-coated layer. To illustrate the point consider Figure 27, which is a block diagram of a fictional coating track.

In this fictional example, the unwinder has a take-up magazine so that parent rolls of base can be spliced onto a continuously moving web. It also has three coating stations, where solutions can be coated on the moving base (known as a web) and three dryers, where the solutions can be dried prior to the next coating station. The reeler also has a take-up magazine so that the fully coated roll may be removed from the machine, without stopping the coating process. Each of the three coating stations has four delivery kettles each capable of replenishment from further kettles, where solution preparation can take place. In this particular example – assuming that one kettle delivers one unique solution – coating machine is only capable of delivering, coating and drying 12 layers. A film structure that requires more layers will need to

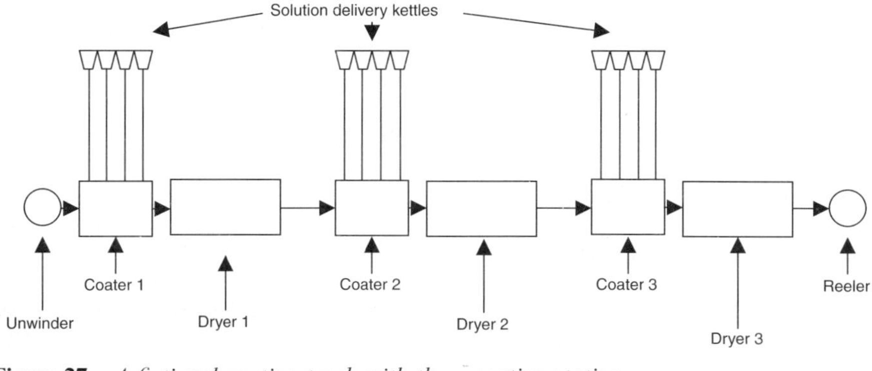

Figure 27 *A fictional coating track with three coating stations*

be partially coated and reeled up. At a later date, the pre-coated web will need to be unwound in the dark and the remaining solutions applied. Although possible, this is not a desired coating method. For example, this coating option might require more track time to coat a product if there are insufficient delivery systems to coat the product in one pass. Two passes, each coating at the same web speed, will require twice the manufacturing time.

In addition to the physical issues associated with coating multiple passes of a coated product, there are also the chemical consequences. For example, should hardener, which is designed to cross-link the gelatin, be applied to the first pass? If so, how much and should matte beads be added so that there are no peel apart static issues when coating the second pass?

Despite these concerns, the coating structure was used to good effect in fully manufactured colour films that were commercially available for over 10 years. The modern trend, however, has been to utilise the advantages of tabular emulsions in order to obtain the relevant red sensitive emulsion speed.

It is highly unlikely that a photographic manufacturer would normally provide details of the composition of one of its products. The patent concerning masking couplers by Crawley *et al.*, mentioned as ref 10 in Chapter 8 and ref 9 in this chapter, provides chemical structures and laydown rates for a colour negative film that was used for the evaluation of novel masking couplers. It is highly unlikely that Crawley *et al.* would have designed a colour negative multilayer for themselves, including all of the components. The more likely scenario is that they used a current formula, using the components, which were already available and then substituted their test materials into the 'product' film formulation for their evaluations.

Accordingly the colour negative film structure described in ref 9 will be examined in more detail as a means of demonstrating all of the chemicals described in all of the earlier chapters. Figure 28 shows the film structure of this film.

In the description of each layer provided below the numbers in the tables refer to coating weights of the various components in gm^{-2}.

9.4.1 Anti-Halation Undercoat Layer

The anti-halation undercoat layer is shown in Figure 29.

9.4.2 Slow Red Sensitive Layer

This layer is comprised of a blend of two red sensitised tabular silver iodobromide emulsions, respectively containing 1.5 M and 4.1 M% iodide, based on silver (Figure 30).

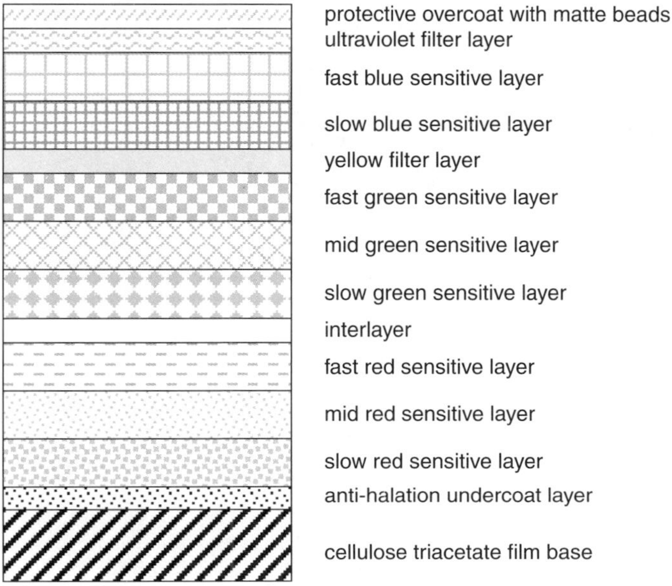

Figure 28 *The layer structure of a colour negative film*

9.4.3 Mid Red Sensitive Layer

This layer is comprised of a red sensitised tabular silver iodobromide emulsion containing 4.1 M% iodide, based on silver (Figure 31).

9.4.4 Fast Red Sensitive Layer

This layer is comprised of a red sensitised tabular silver iodobromide emulsion containing 3.7 M% iodide, based on silver (Figure 32).

9.4.5 Interlayer

For interlayer, see Figure 33.

9.4.6 Slow Green Sensitive Layer

This layer is comprised of a blend of a lower and higher (lower and higher grain ECD) sensitivity, green sensitised tabular silver iodobromide emulsions, respectively containing 2.6 M and 4.1 M% iodide, based on silver (Figure 34).

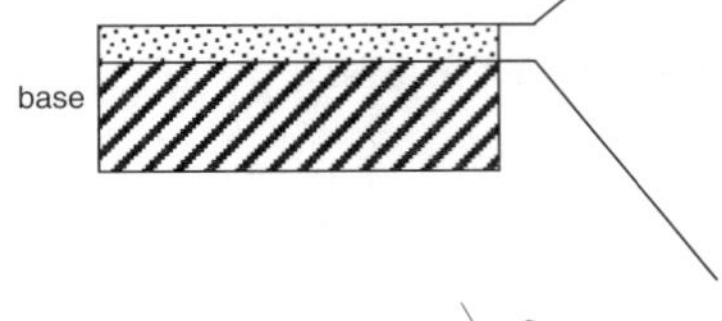

Component	Laydown rate
Black colloidal silver sol	0.107
UV-A	0.075
UV-B	0.075
Oxidised developer scavenger	0.161
Cyan printing dye	0.034
Magenta printing dye	0.013
Yellow printing dye	0.095
Tritoluoyl phosphate	0.105
Di-n-butyl phthalate	0.399
Tris(2-ethyl)phosphate	0.013
Disodium salt of 3,5-disulphocatechol	0.215
Gelatin	2.152

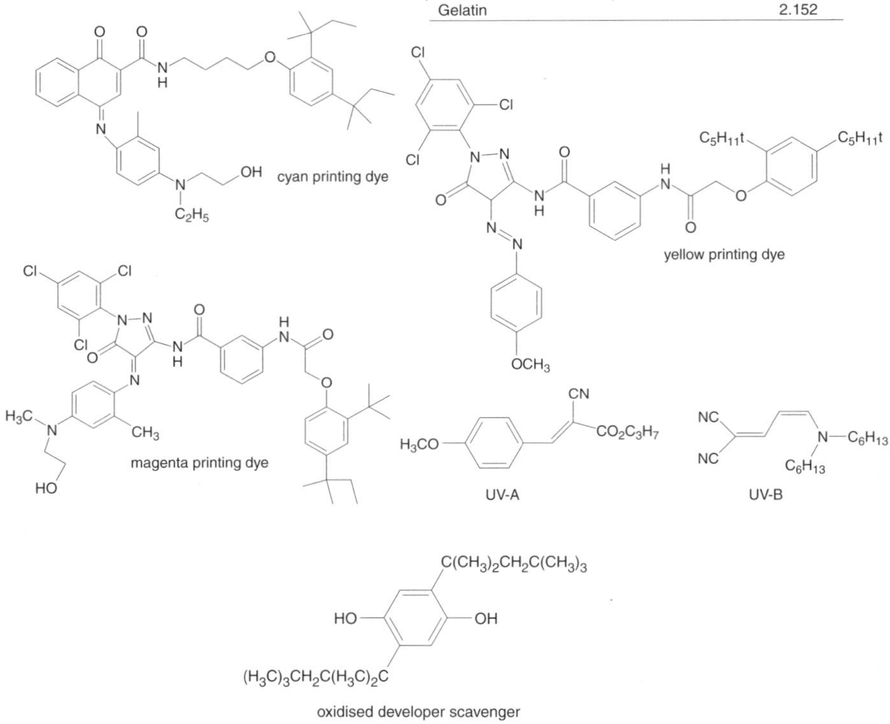

Figure 29 *Components of the anti-halation undercoat layer*

9.4.7 Mid Green Sensitive Layer

This layer is comprised of a blend of two emulsions, which are green sensitised tabular silver iodobromide emulsions each containing 4.1 M% iodide, based on silver (Figure 35).

9.4.8 Fast Green Sensitive Layer

This layer is comprised of a green sensitised tabular silver iodobromide emulsion containing 4.1 M% iodide, based on silver (Figure 36).

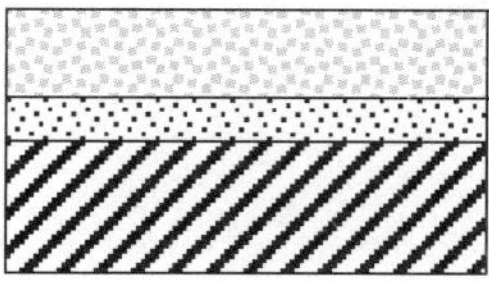

Component	Laydown rate
AgIBr (0.55 gm ECD, 0.08 gm t)	0.355
AgIBr (0.66' cm ECD, 0.12 gm t)	0.328
Bleach accelerator coupler	0.075
Cyan dye forming DIR coupler	0.015
Cyan dye forming coupler	0.359
N-n-Butyl acetanilide	0.030
N,N-Diethyl lauramide	0.098
4-Hydroxy-6-methyl-1,3,3a,7-tetraazaindine Na	0.011
Gelatin	1.668

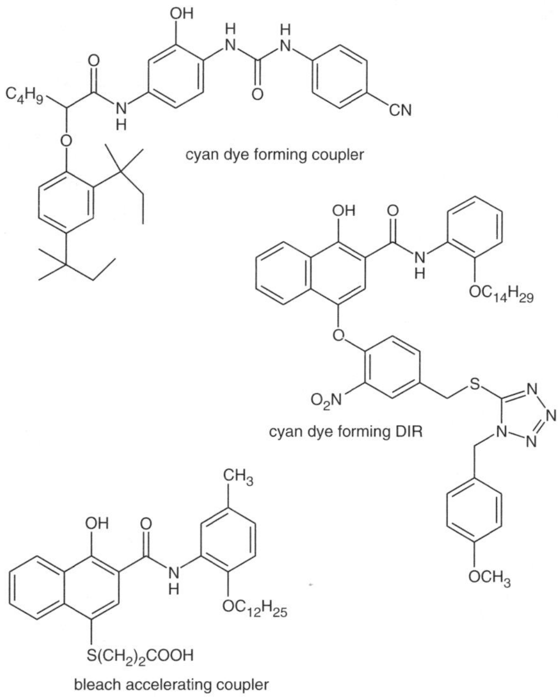

Figure 30 *Components of the slow red sensitive layer*

9.4.9 Yellow Filter Layer

For yellow filter layer, see Figure 37.

9.4.10 Slow Blue Sensitive Layer

This layer is comprised of a blend of three blue sensitised tabular silver iodobromide emulsions, respectively containing 1.5, 1.5 and 4.1 M% iodide, based on silver (Figure 38).

The formula provided in the patent by Crawley *et al.*[9] also lists a cyan dye forming DIR in his layer. It would be more common to use a yellow

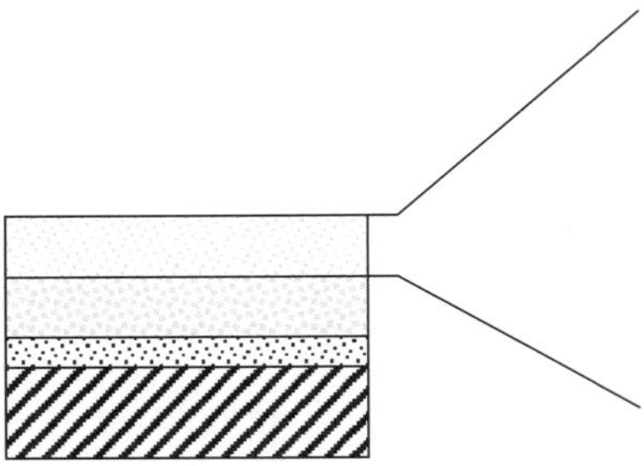

Component	Laydown rate
AgIBr (1.30 gm ECD, 0.12 um t)	1.162
Bleach accelerator coupler	0.005
Cyan dye forming DIR coupler	0.016
Cyan dye forming magenta coloured coupler	0.059
Cyan dye forming coupler	0.207
Di-n-butyl phthalate	0.207
N-n-Butyl acetanilide	0.032
N,N-Diethyl lauramide	0.007
4-Hydroxy-6-methyl-1,3,3a,7-tetraazaindine Na	0.019
Gelatin	1.291

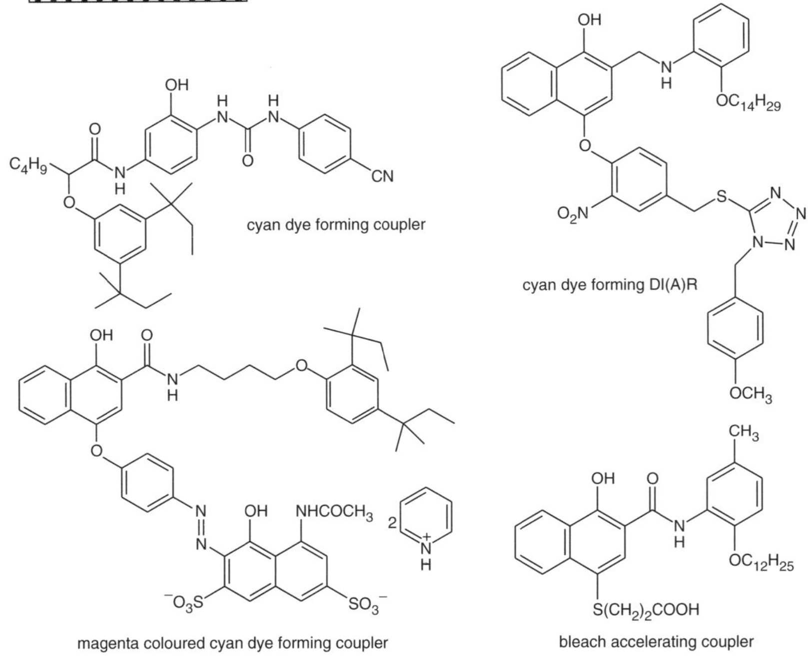

Figure 31 *Components of the mid red sensitive layer*

dye forming coupler in this blue sensitised layer. This cyan dye forming DIR coupler has therefore been omitted from this account of Crawley's formulation.

9.4.11 Fast Blue Sensitive Layer

This layer is comprised of a blue sensitised silver iodobromide emulsion containing 9.0 M% iodide, based on silver (Figure 39).

9.4.12 Ultraviolet Filter Layer

For ultraviolet filter layer, see Figure 40.

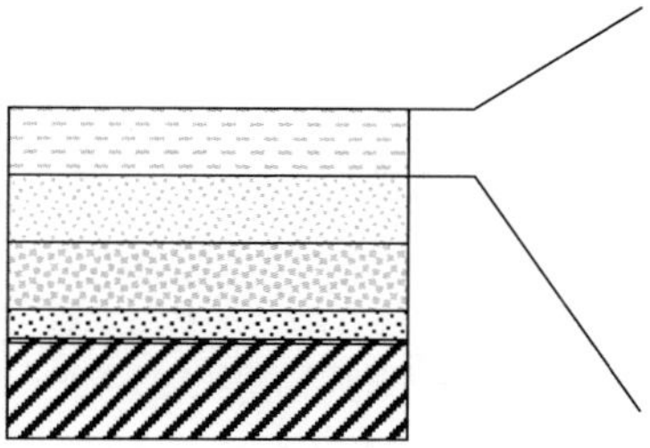

Component	Laydown rate
AgIBr (2.61*gm* ECD, 0.12 gm t)	1.060
Bleach accelerator coupler	0.005
Cyan dye forming DIAR	0.048
Cyan dye forming DIR	0.027
Cyan dye forming magenta coloured coupler	0.022
Cyan dye forming coupler	0.312
Tritoluoyl phosphate	0.194
Di-n-butyl phthalate	0.274
N-n-Butyl acetanilide	0.054
N,N-Diethyl lauramide	0.007
4-Hydroxy-6-methyl-1,3,3a,7-tetraazaindine Na	0.010
Gelatin	1.291

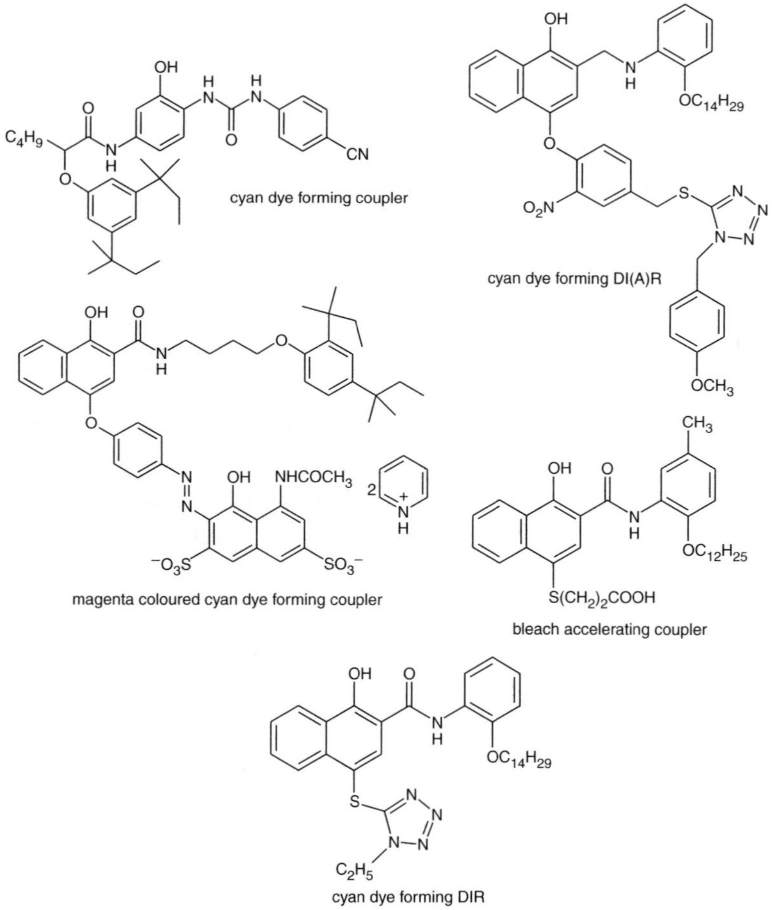

Figure 32 *Components of the fast red sensitive layer*

9.4.13 Supercoat (Protective Overcoat) Layer

The laydown rates of the printing dyes in the AHU layer can be adjusted if there is a source of variability, in the rest of the formulation. There is another useful feature, however, that relates to printing families of films

Component	Laydown rate
Oxidised developer scavenger	0.086
Tris(2-ethylhexyl)phosphate	0.129
Gelatin	0.538

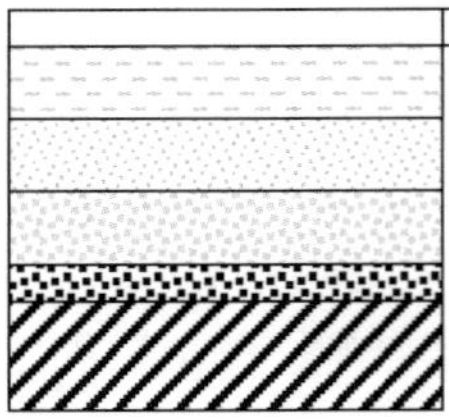

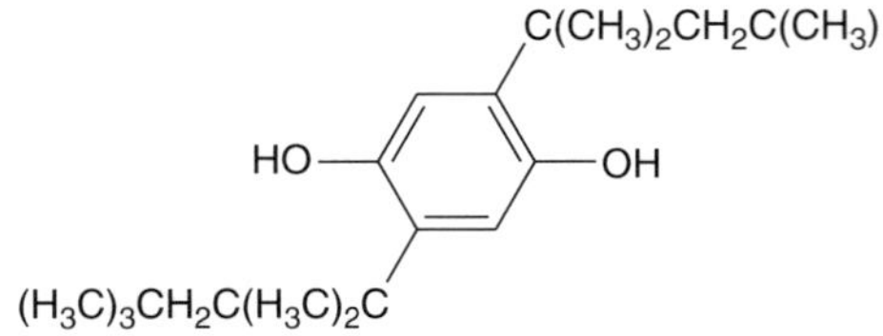

oxidised developer scavenger

Figure 33 *Components of the interlayer*

on the same printer setting. Different film products may require different laydown rates of coloured masking couplers, *etc.* that would result in different products from the same manufacturer needing a different printer setting in order to adjust for differences in the film density at zero exposure. The ability to change the laydown rate of the printing dyes in the respective products allows the printing settings to remain constant and the films to be of different colour balance, or indeed to some extent photographic speed (Figure 41).

The use of bleach accelerating releasing couplers is to ensure that there is no silver or any derivatives remaining in the coating after processing. The mechanism for the release of the bleach accelerating moiety is identical to that of other couplers.

9.5 Graphic Arts Film

Graphic arts films were used extensively in the printing industry prior to the advent of computer to plate technology. Film formats are much larger than amateur camera film formats and could be as large as 1 m on their largest dimension. The films are exposed using image setting lasers which could be of a variety of wavelengths of light. At their peak there were well over 100 different commercially available films, although some were merely size format changes using identical chemistry. Given the extent of the variety of formats that were available, the discussion here will centre on the chemistry that was common to the various films.

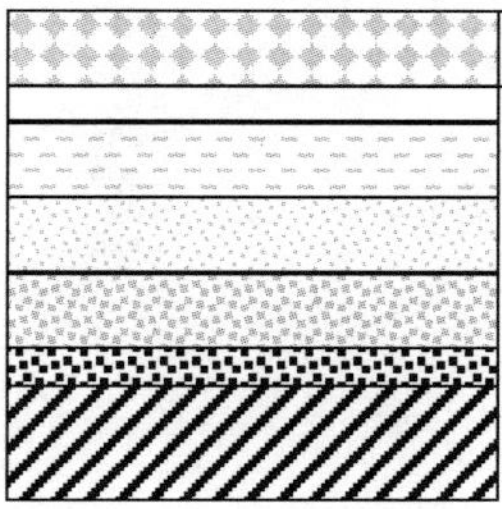

Component	Laydown rate
AgIBr (0.81 gm ECD, 0.12 gm t)	0.251
AgIBr (0.92 gm ECD, 0.12 gm t)	0.110
Magenta dye forming yellow coloured coupler	0.070
Magenta dye forming coupler	0.339
Stabiliser	0.034
Tritoluoyl phosphate	0.305
4-Hydroxy-6-methyl-1,3,3a,7-tetraazaindine Na	0.006
Gelatin	1.721

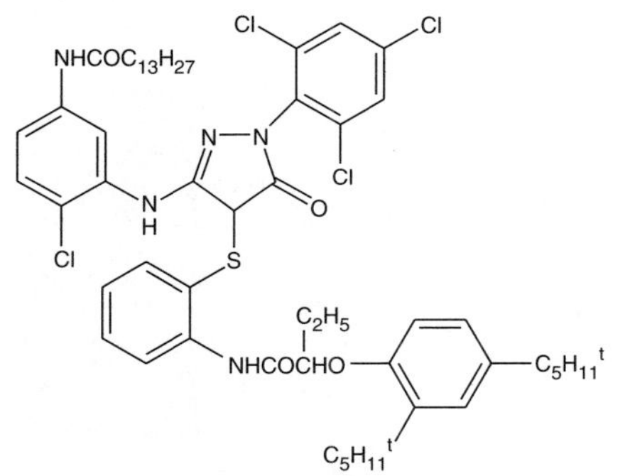

magenta dye forming coupler

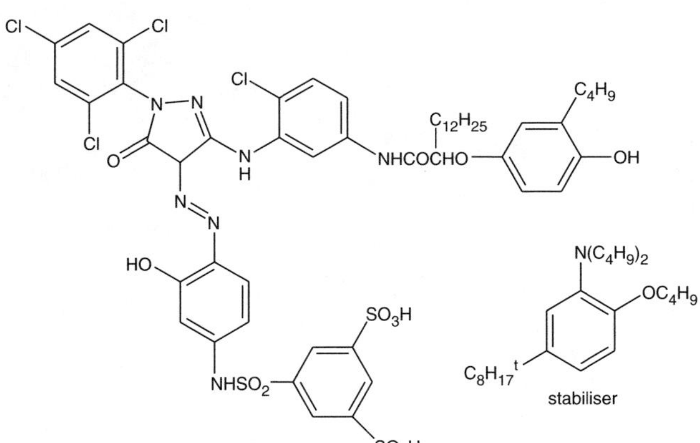

stabiliser

an experimental yellow coloured magenta coupler

Figure 34 *Components of the slow green sensitive layer*

A generalised film format appears as shown in Figure 42.

Sheet formats of any size will be subject to stresses on the side bearing the gelatin coating. A coating on the other side of the base is therefore applied in order to balance out these physical stresses. The base is usually cast polyethylene terephthalate, which can be of a number of

Component	Laydown rate
AgIBr (0.92 gm ECD, 0.12 gm t)	0.113
AgIBr (1.22 gm ECD, 0.11 gm t)	1.334
Development inhibitor	0.032
Magenta dye forming yellow coloured coupler	0.154
Magenta dye forming coupler	0.087
Oxidised developer scavenger	0.018
Tritoluoyl phosphate	0.079
Di-n-butyl phthalate	0.032
Stabiliser	0.009
4-Hydroxy-6-methyl-1,3,3a,7-tetraazaindine Na	0.023
Gelatin	1.668

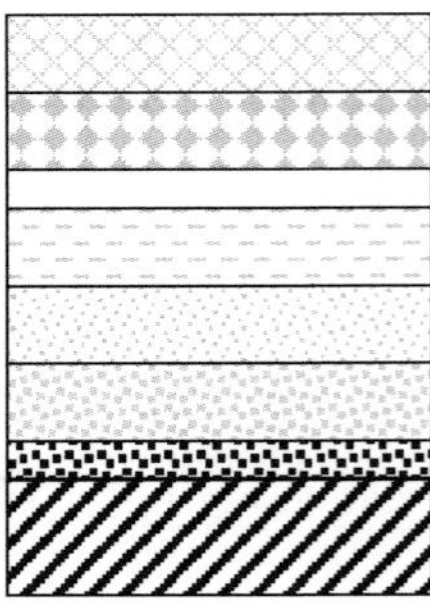

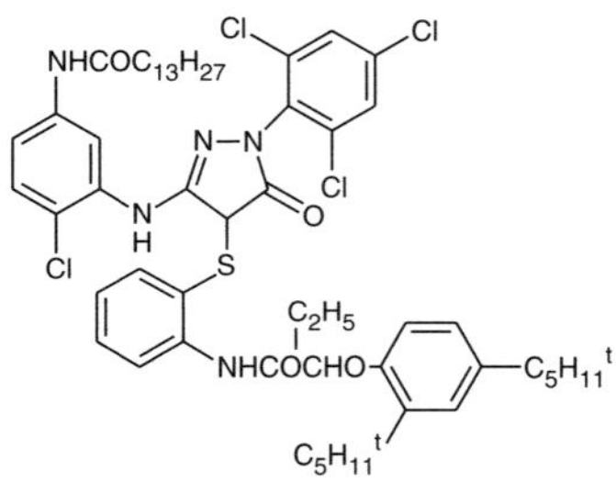

magenta dye forming coupler

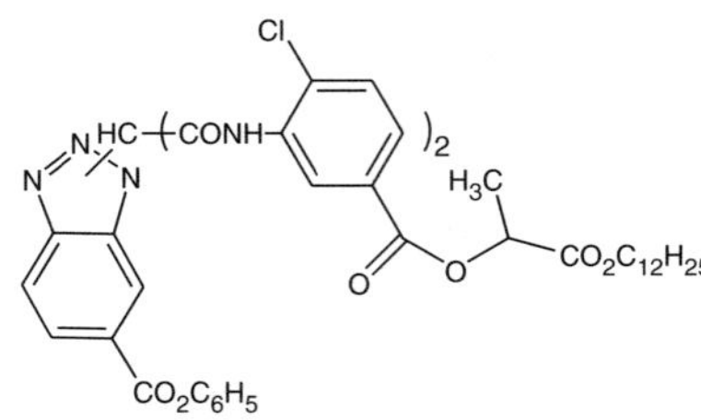

development inhibitor

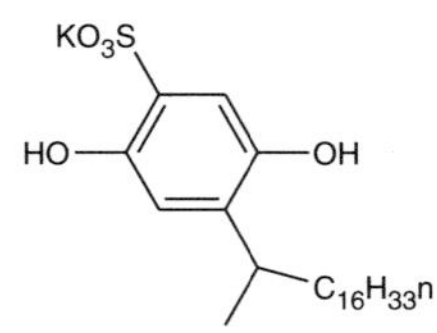

oxidised developer scavenger

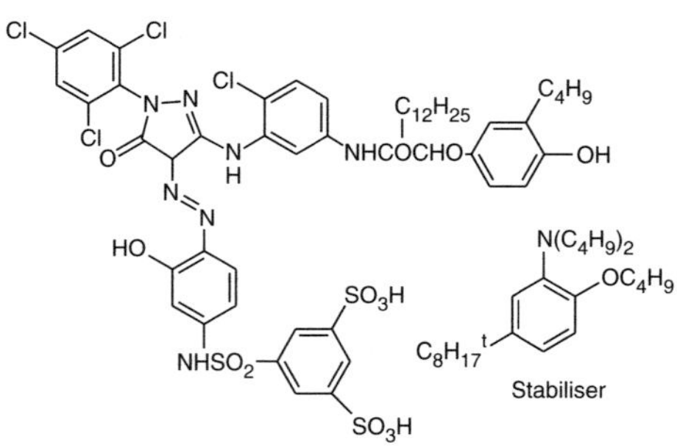

an experimental yellow coloured magenta coupler

Stabiliser

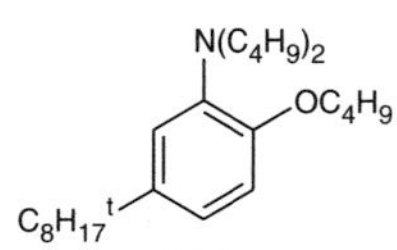

stabiliser

Figure 35 *Components of the mid green sensitive layer*

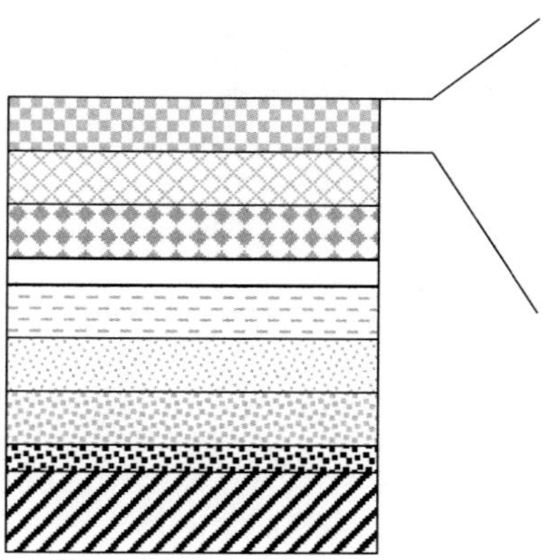

Component	Laydown rate
AgIBr (2.49 gm ECD, 0.14 gm t)	0.909
Magenta dye forming DIR	0.003
Magenta dye forming DIAR	0.032
Magenta dye forming yellow coloured coupler	0.070
Magenta dye forming coupler	0.113
Tritoluoyl phosphate	0.108
Di-n-butyl phthalate	0.065
Stabiliser	0.011
4-Hydroxy-6-methyl-1,3,3a,7-tetraazaindine Na	0.011
Gelatin	1.405

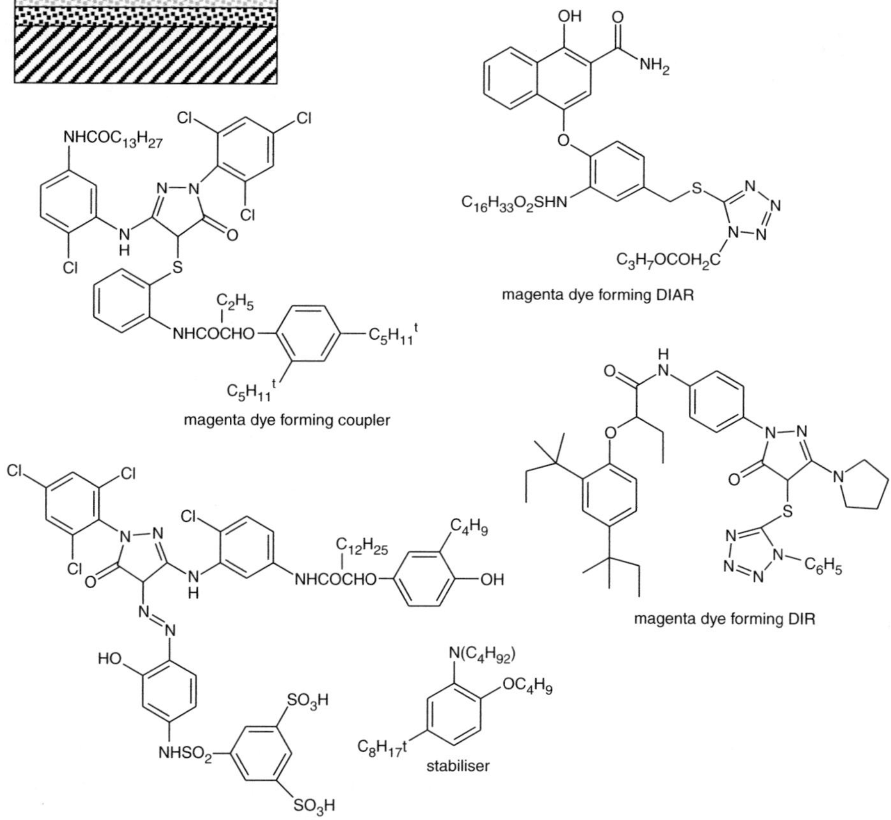

Figure 36 *Components of the fast green sensitive layer*

thicknesses. This layer is also used to coat a dye the purpose of which is to prevent halation in a similar way to that of an anti-halation undercoat layer mentioned previously, Figure 29. The dyes used are all organic dyes and vary depending upon the product use and wavelength of the exposing laser.

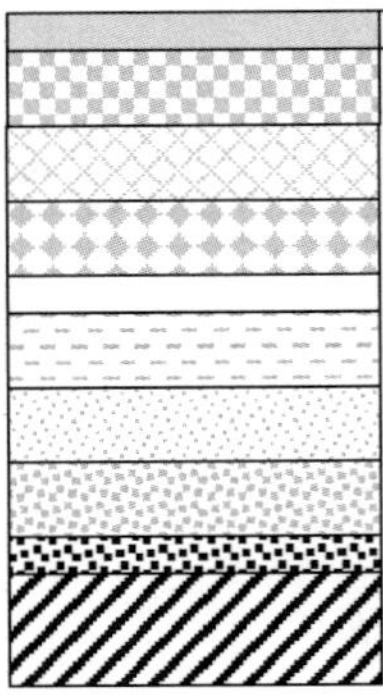

Component	Laydown rate
Oxidised developer scavenger	0.086
Tris(2-ethylhexyl)phosphate	0.129
Gelatin	0.646

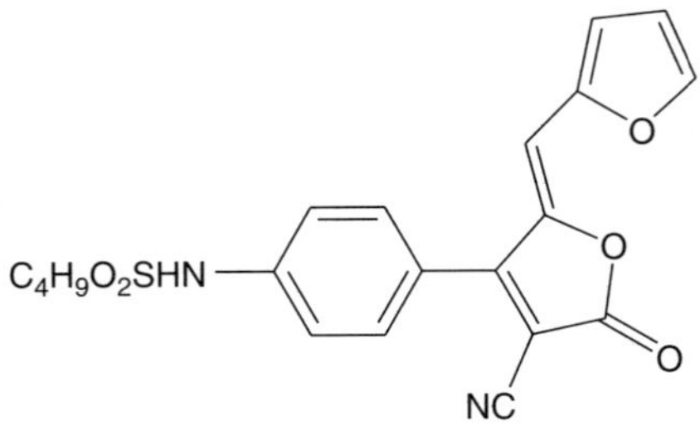

yellow filter dye

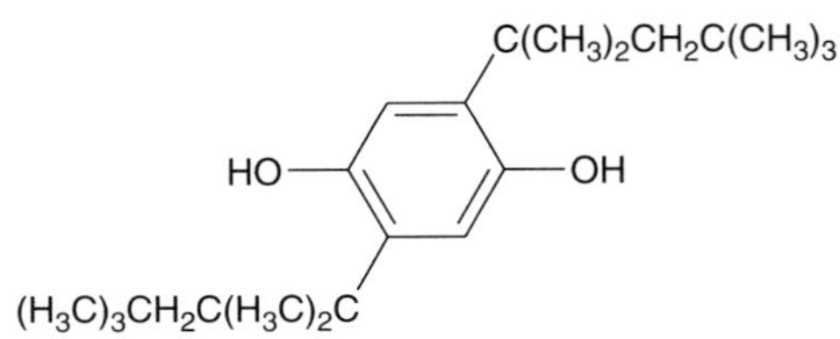

oxidised developer scavenger

Figure 37 *Components of the yellow filter layer*

The density log exposure curve for most if not all graphic arts products is similar in that each film has a very high maximum density, often of the order of 6 density units, see Figure 43.

The slope of the plot often approaches 5 or perhaps 5.5 compared with 0.7 or so for the colour negative films. The need for this photographic response has been covered in Chapter 1, Figures 18, 19 and 20, where the concept of half-tone dots was examined.

The use of lasers as the means of exposing the films has led to a range of emulsion sensitising dyes that have not thus far been reported. In their 2002 patent Gray *et al.*[21] reported the use of a mixture of cubic monodispersed emulsions, one being a 70:30 chlorobromide of edge length 0.21 μm, and the other being a chlorobromide emulsion of edge

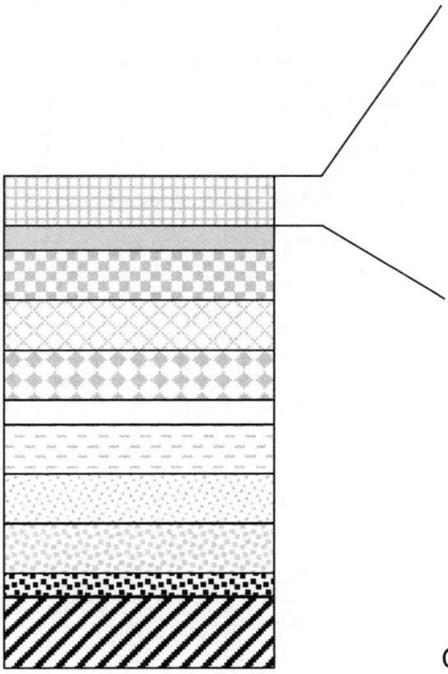

Component	Laydown rate
AgIBr (0.55 gm ECD, 0.08 gm t)	0.156
AgIBr (0.77 gm ECD, 0.14 gm t)	0.269
AgIBr (1.25 gm ECD, 0.14 gm t)	0.430
Yellow dye forming DIR	0.054
Yellow dye forming coupler	1.022
Bleach accelerator coupler	0.011
Tritoluoyl phosphate	0.538
N-n-Butyl acetanilide	0.054
N,N-Diethyl lauramide	0.014
4-Hydroxy-6-methyl-1,3,3a,7-tetraazaindine Na	0.014
Gelatin	2.119

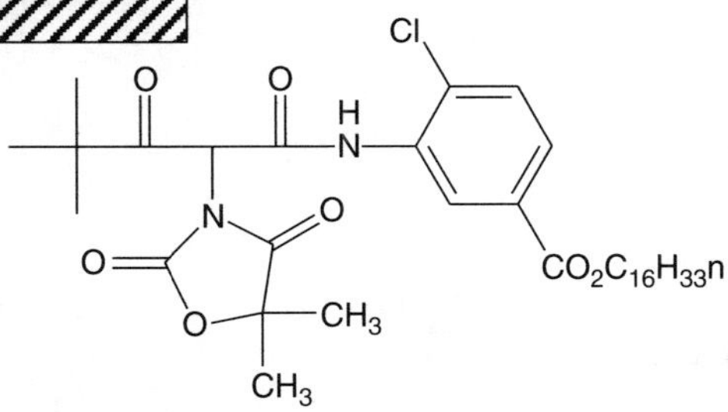

yellow dye forming coupler

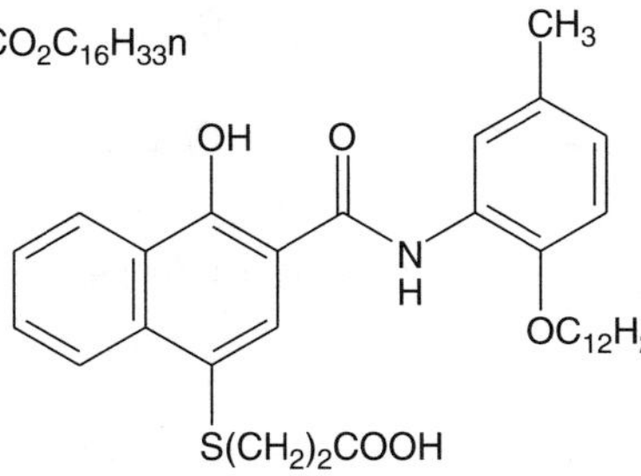

bleach accelerating coupler

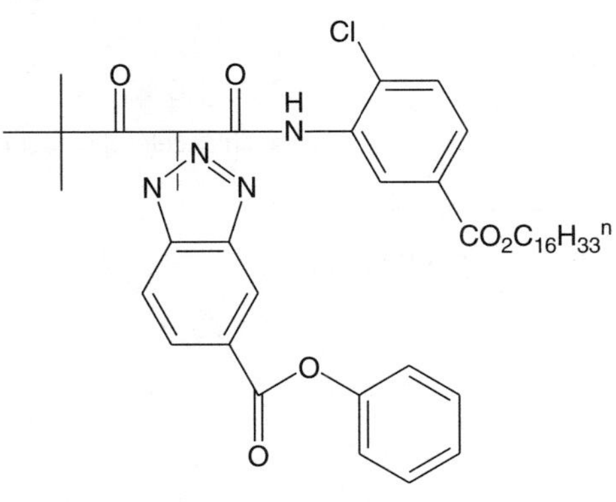

yellow dye forming DIR coupler

Figure 38 *Components of the slow blue sensitive layer*

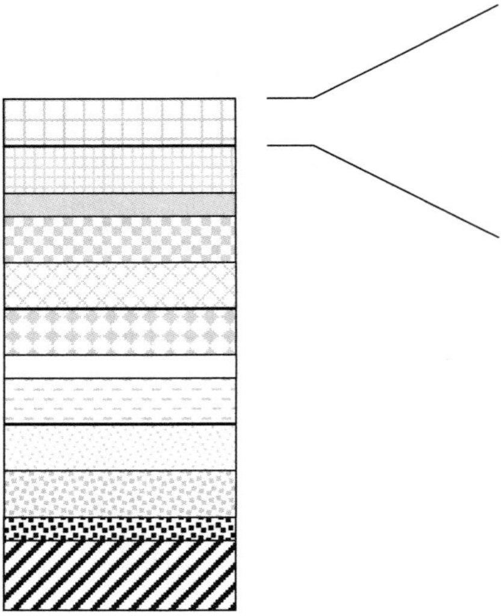

Component	Laydown rate
AgIBr (1.04 gm ECD)	0.699
Unsensitised silver bromide Lippmann emulsion	0.054
Yellow dye forming coupler	0.473
Yellow dye forming DIR coupler	0.086
Bleach accelerator coupler	0.005
Tritoluoyl phosphate	0.280
N,N-Diethyl lauramide	0.004
4-Hydroxy-6-methyl-1,3,3a,7-tetraazaindine Na	0.012
Gelatin	1.183

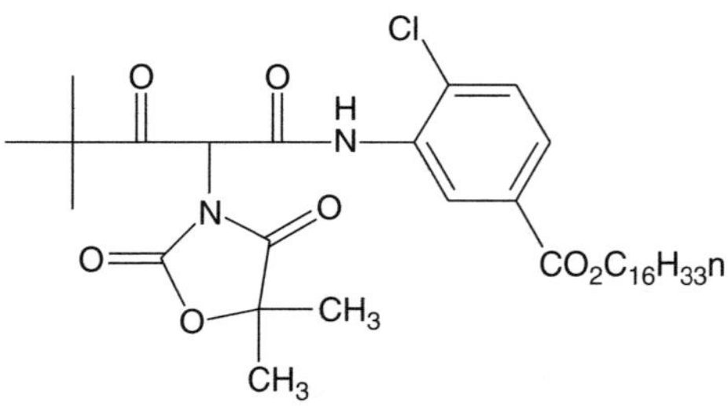

yellow dye forming coupler

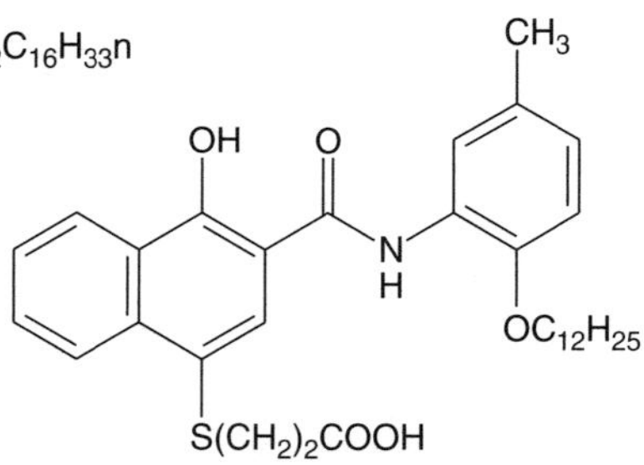

bleach accelerating coupler

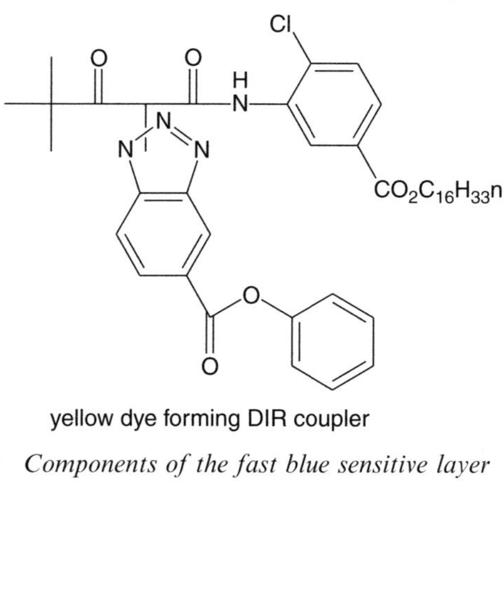

yellow dye forming DIR coupler

Figure 39 *Components of the fast blue sensitive layer*

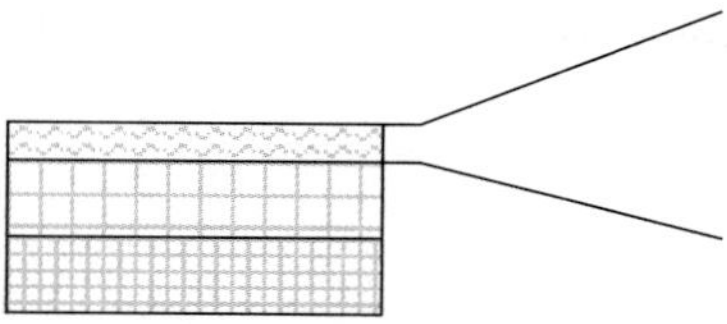

Component	Laydown rate
UV-A	0.108
UV-B	0.108
Unsensitised silver bromide Lippmann emulsion	0.215
Tritoluoyl phosphate	0.151
Gelatin	0.699

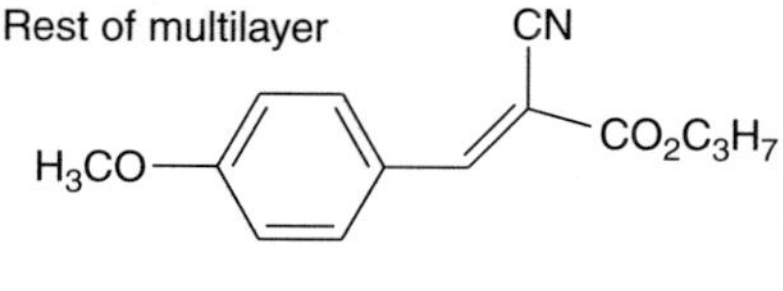

UV-A

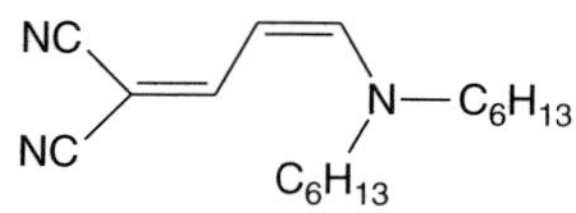

UV-B

Figure 40 *Components of the ultraviolet filter layer*

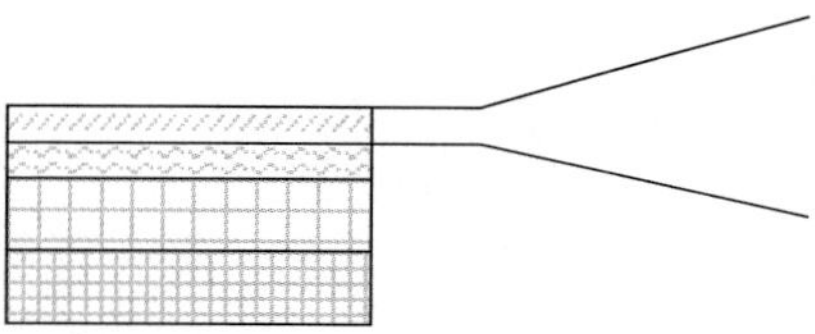

Component	Laydown rate
Polymethylmethacrylate matte beads	0.005
Soluble polymethylmethacrylate matte beads	0.108
Silicone lubricant	0.039
Gelatin	0.882

Figure 41 *Components of the supercoat layer*

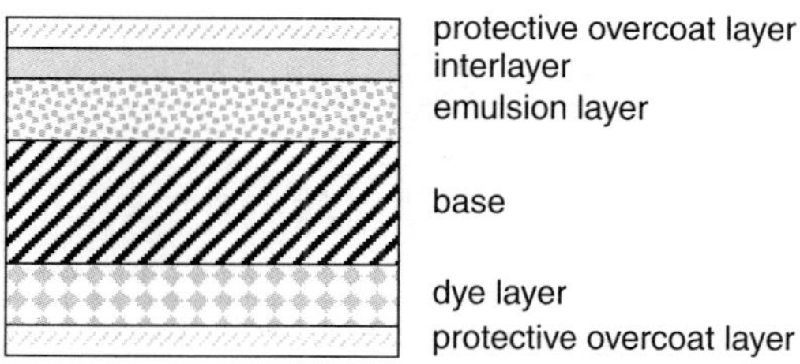

Figure 42 *A typical graphic arts layer structure*

length 0.18 μm. Figure 44 provides examples of the typical dyes used to adsorb onto the silver halide grain surface.

Gray *et al.*[21] suggest that R^8, R^9 and R^{10} can represent an alkyl group which may be substituted for example with acid water-solubilising groups, R^{11} and R^{12} can be an alkyl group of 1-4 carbon atoms, R^{13}, R^{14} and R^{15} represent substituted or unsubstituted aryl moieties and X is a halogen.

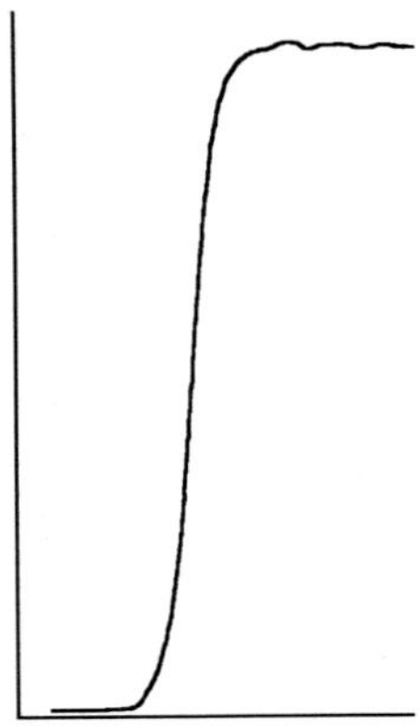

Figure 43 *Density vs log exposure plot for a typical graphic arts film*

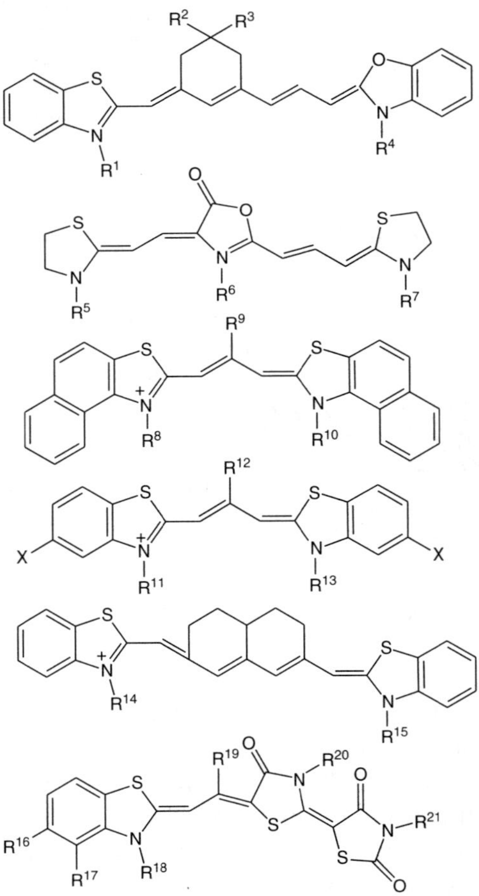

Figure 44 *Typical graphic arts sensitising dyes*

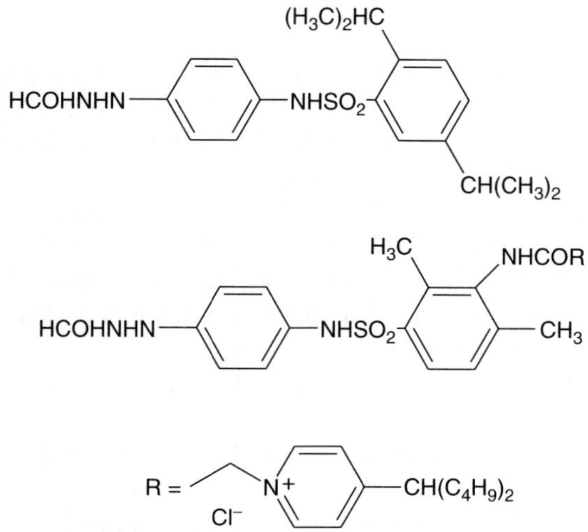

Figure 45 *Nucleating agents used in graphic arts films*

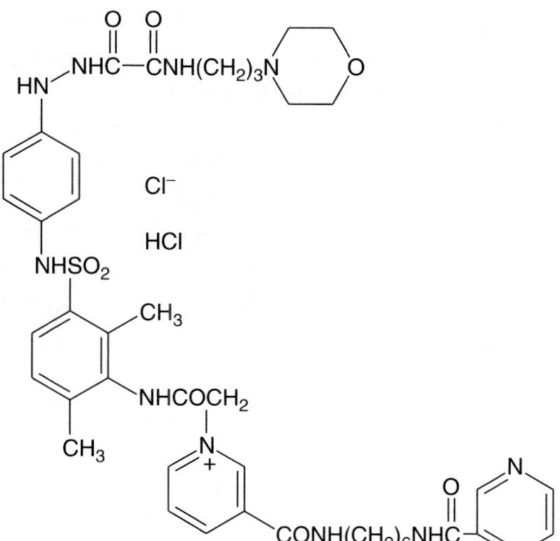

Figure 46 *An example of a more recently used nucleating agent*

The high contrast of the films is usually achieved by the use of nucleating agents. These nucleating agents are often hydrazides, see for example Figure 45.[22]

More recently, Coldrick *et al.*[23] reported on the use of compounds similar to Figure 46 as nucleating agents.

The emulsion halide content and the type of product are factors that help to determine which nucleating agent is used for which film.

References

1. A.H. Tsou, J. Greener, G.D. Smith and G.M. Mosehauer, US 5,607,826, Eastman Kodak Co.
2. R. Khanna and F.J. Jacoby, US 3,919,156, Eastman Kodak Co.
3. J. Greener and Wen-Li A Chen, US 5,599,658, Eastman Kodak Co.
4. E.E. Arrington and R.J. Kehl, US 5,529,737, Eastman Kodak Co.
5. K.N. Kilminster and D. Hoke, US 4,775,616, Eastman Kodak Co.
6. I. Masakuni and I. Isaburo, GB 1,269,788, Konishiroku Photo Industry.
7. A.R. Pitt, B.A. Clark and J.F. Padday, US 5,366,857, Eastamn Kodak Co.
8. GB 1,261,156, Fuji Photo Film Co. Ltd.
9. J.N. Younathan, M.W. Crawley and K. Chari, US 6,132,943, Eastman Kodak Co.
10. M. Slagt, A. Kase, P van Asten and Y. Iwasa, WO 2004-081661, Fuji Film B.V.
11. I. Tsumoru and S. Shinji, US 4,396,706, Fuji Photo Film Co. Ltd.
12. O. Hisashi and M. Yasuo, US 5,057,407, Fuji Photo Film Co. Ltd.
13. D. Smith and J.L. Muehlbauer, US 5,378,577, Eastman Kodak Co.
14. A. Eeles and A. O'Neill, GB 1,500,497, Kodak Limited.
15. A. Weissberger and P.W. Vittum, U.S. 2,474,293, Eastman Kodak Co.
16. I.F. Salminen and C.R. Barr, U.S. 2,895,826, Eastman Kodak Co.
17. A. Loria, U.S. 3,476,563, Eastman Kodak Co.
18. A. Loria and P.W. Vittum, U.S. 2,600,788, Eastman Kodak Co.
19. G.J. Lestina, U.S. 3,519,429, Eastman Kodak Co.
20. L.E. Beavers, U.S. 2,983,608, Eastman Kodak Co.
21. C.J. Gray, A.R. Benoy and E. Ordia, US 6,372,417, Eastman Kodak Co.
22. P.J. Coldrick and J. Pich, US 5,342,732, Eastman Kodak Co.
23. J.A. Bogie, P.J. Coldrick, J.D. Goddard and L.J. Leyshon, US 6,245,480, Eastman Kodak Co.

CHAPTER 10

Paper Structures

If left untreated, a paper base will absorb moisture and be ineffective as a support for photographic products. A polymer layer is therefore extruded onto the paper support. During the same process the relevant company logo is printed on the reverse of the paper. A typical extrusion process was recently published by Fuji Photo Film B.V.,[1] a schematic diagram of the process for which appears as Figure 1.

Several different compounds are reported in the patent, for example ref 1

- the blue dye fastusol was added to counteract the effects of paper yellowing,
- a mixture of 65 wt% calcium carbonate (1–2 μm particle size), 15 wt% of calcium carbonate (<1.0 μm particle size) and 20 wt% styrene butadiene,
- kaolin (15 wt%) replaced the finer calcium carbonate particles in one example, and

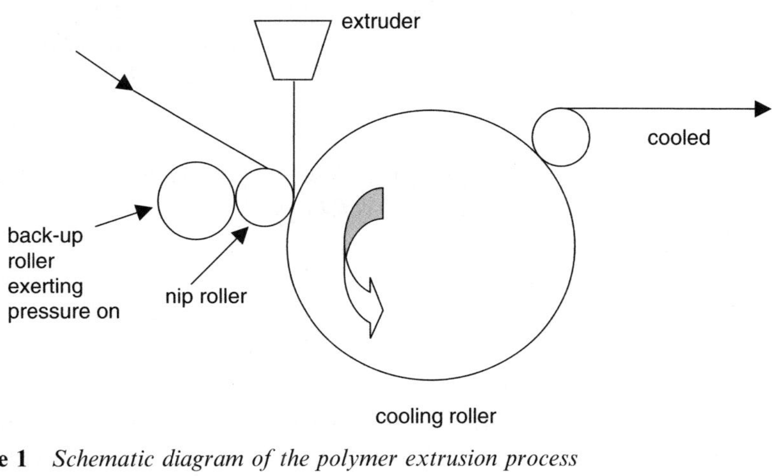

Figure 1 *Schematic diagram of the polymer extrusion process*

- a methyl methacrylate-butadiene latex, polyacrylate latex, polyvinyl alcohol or polysaccharides were also reported as variants in the polymer either in combination or used separately.

In their 2001 European patent Bouwstra and Vries[2] further describe the process of extrusion, particularly the reduced number of defects associated with using corona discharge during the process. They comment (in claim 27)

> *...Before extrusion-coating of the topside, the substrate was pretreated with corona and subsequently preheated with NIR radiation up to a temperature of 39°C or 83°C. The treating width of the NIR-equipment covered 24 cm in the centre of the total web width of 47 cm. For the pretreatment two NIR-units were used. After the online pretreatments (Corona and preheating) the substrate was extrusion coated at the topside with a polyolefin resin LDPE of 28 g/m^2 (containing metaloxide, dyes, coloured pigments, optical brighteners and the like) at a speed of 400 m/min. The hotmelt (temperature 326°C) was nipped (pressure 6 bar) between the substrate and a cooling chill-roll. Because of the NIR-pretreatment the crater-like defects or so-called 'pits' defects decreased in comparison to the samples which received no heat pretreatment (remaining at room temperature). The NIR-pretreatment shows no disadvantage in the number of releasing defects, when the extrusion coated web releases from the cooling chill-roll.*

The results of this pre-treatment were shown in Figure 2.

Many polymers have been extruded onto paper base by all of the photographic manufacturing companies. In one of his patents, Edwards of the Eastman Kodak Co. describes a number of polymers,[3] for example

> *...In accordance with this invention, polyolefin extrusion coating compositions are provided which provide coatings having good coatability and provide a coating having good adhesion to the substrate. Such compositions are a blend of*

- *at least one crystalline polypropylene or propylene containing copolymer containing at least 80 weight percent propylene,*
- *low density polyethylene,*
- *tackifying resin, and*
- *fatty acid amide.*

These extrusion coating compositions provide an extrusion coating composition having a melt flow rate of 40 to 100 that can be coated at coating speeds of greater than 122 m/min. to provide coatings of less than 1

	Paper type 1		Paper type 2	
	Number of pits per 4mm^2		Number of pits per 4mm^2	
	(800-1200 jm^2)	(>1200 jm^2)	(800-1200 μm^2)	(>1200 μm^2)
No NIR treatment(23°C)	9.3	2.3	22.3	9.3
NIR treatment (39°C)	1.7	0.3	14.3	5.0
NIR treatment (83°C)	2.0	1.3	9.7	1.3

Figure 2 *NIR pre-treatment results*

	Weight Percent Contained in Composition	
Component	Broad Range	Preferred Range
Propylene Homo-or	90-65%	80-90%
Copolymer Low Density	5-20%	5-10%
Polyethylene Tackifier Resin	0.5-12.5%	2-10%
Fatty Acid Amide	0.2-1.0%	0.3-0.5%

Figure 3 *Some of the polymer compositions from Eastman Kodak Co.*

mil thickness and have good adhesion to the substrate and good heat seal strength. These unique blend compositions with which this invention is concerned are as follows:(Figure 3)

Diaz and Mears discuss the tackifier resin and a range of polymers including blends of polypropylene and polyethylene. A crystalline propylene-alpha-monoolefinic block copolymer is also described[4] and the use of DAC-B resin discussed. This resin is a complex mixture of saturated and unsaturated hydrocarbons obtained from the thermal cracking of a hydrocarbon stream to produce ethylene and/or propylene.

10.1 Colour Paper

The previously described technology of slide hopper coating works best at lower coating speeds than is used to coat colour paper. Accordingly, an alternative coating method was reported and is believed to be in use.[5–8] Figure 4 shows a close-up of the hopper with two solutions in order to demonstrate the principle. Figure 5 shows a side view of the falling curtain onto the web, in this case just one solution.

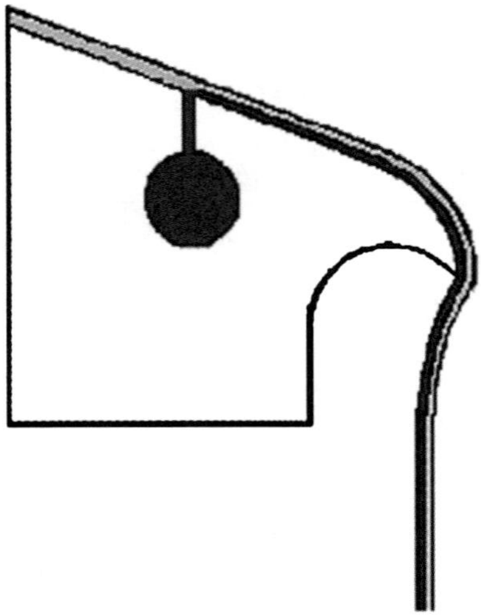

Figure 4 *Schematic diagram of a curtain coating hopper*

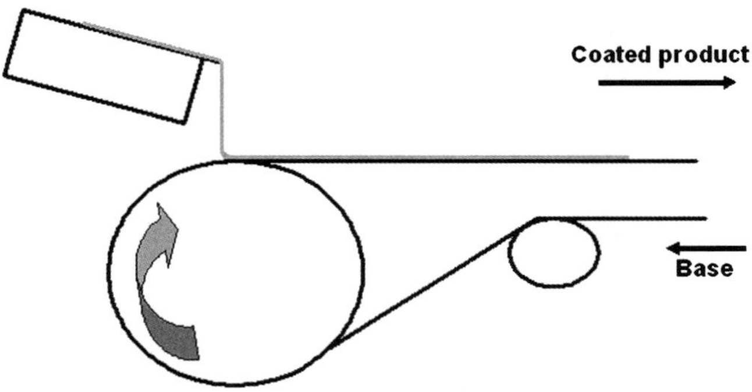

Figure 5 *The relative positions of hopper, moving base and liquid curtain used in curtain coating*

An example of the resulting sensitometry obtained from these and similar coatings is outlined in Figure 6.

The contrast from this paper is intermediate between film and graphic arts film and paper, although closer to that of the graphic arts films. Also, there is no image at zero dye density as a pure white is required at zero exposure. In this example a tail off of the yellow dye at high densities is not an issue as the eye has least sensitivity to blue light (which is being recorded by the yellow dye).

The layer structure for paper is outlined in Figure 7. This example is from the 2004 Fuji Photo Film B.V. patent.[9] This particular colour paper has been formulated for printing from a digital camera and is likely to be a formula that will be prevalent for some time to come.

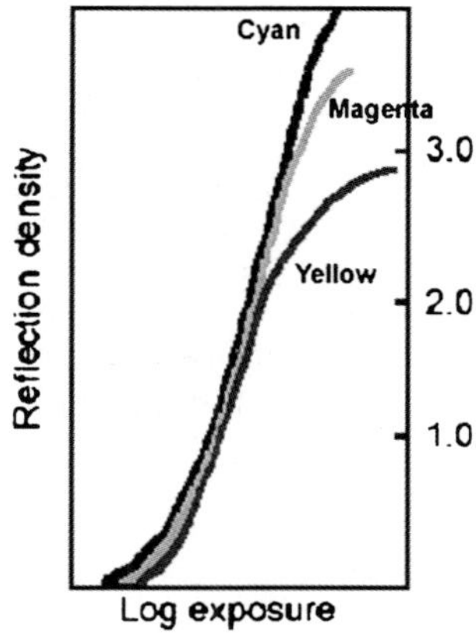

Figure 6 *A density vs log exposure plot for colour paper*

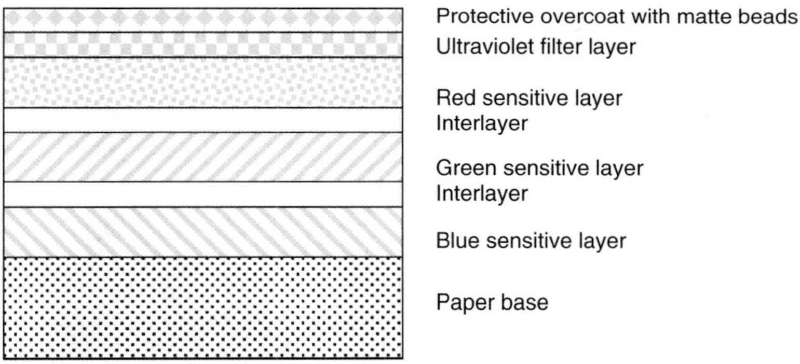

Figure 7 *Layer order for a typical photographic paper*

Throughout this description of the recent colour paper from the Fuji Film Co., all laydown rates are in gm^{-2}. The raw paper was laminated using a corona discharge on both sides with polyethylene. A gel-subbing layer was coated on the top of the coating side polyethylene layer. This gel-subbing layer contained 3 mg m^{-2} of compound A, 12 mg m^{-2} of compound B and 14% by mass of titanium oxide.

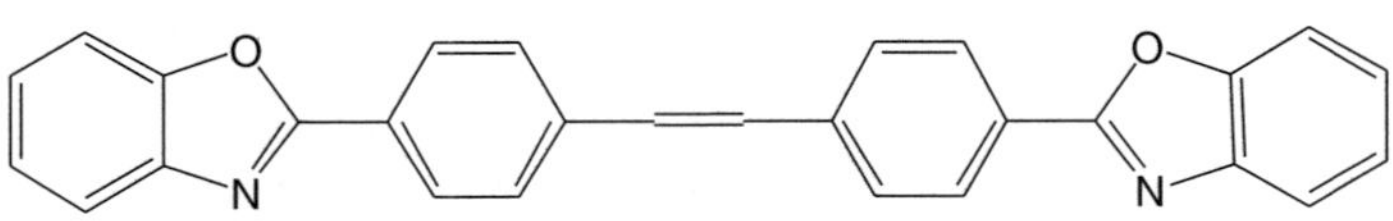

A

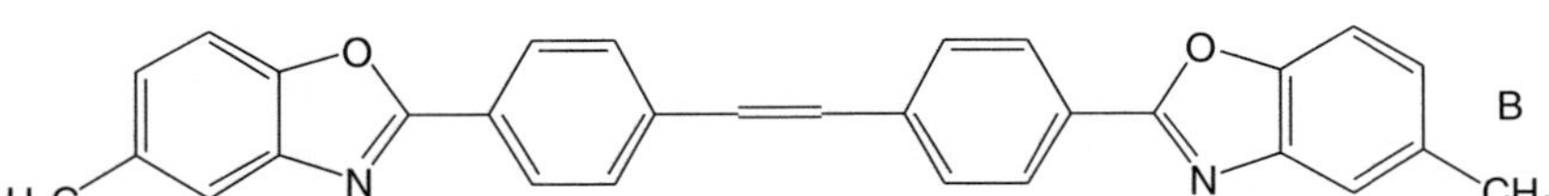

B

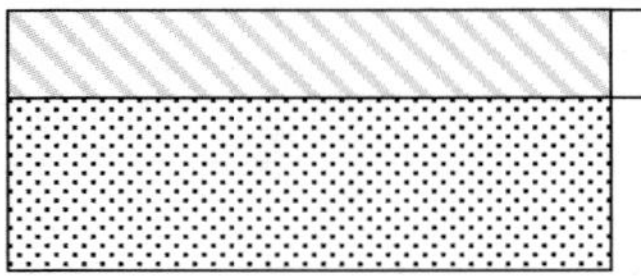

Component	Laydown rate
Gelatin	1.35
Yellow coupler 1	0.41
Yellow coupler 2	0.21
Colour-image stabiliser 1	0.08
Colour-image stabiliser 2	0.04
Colour-image stabiliser 3	0.08
1-(3-methylureidophenyl)-5-mercaptotetrazole	3.3×10^{-4} mol*
4-hydroxy-6-methyl-1,3,3a,7-tetrazaindene	1×10^{-4} mol*

* per mole of the silver halide

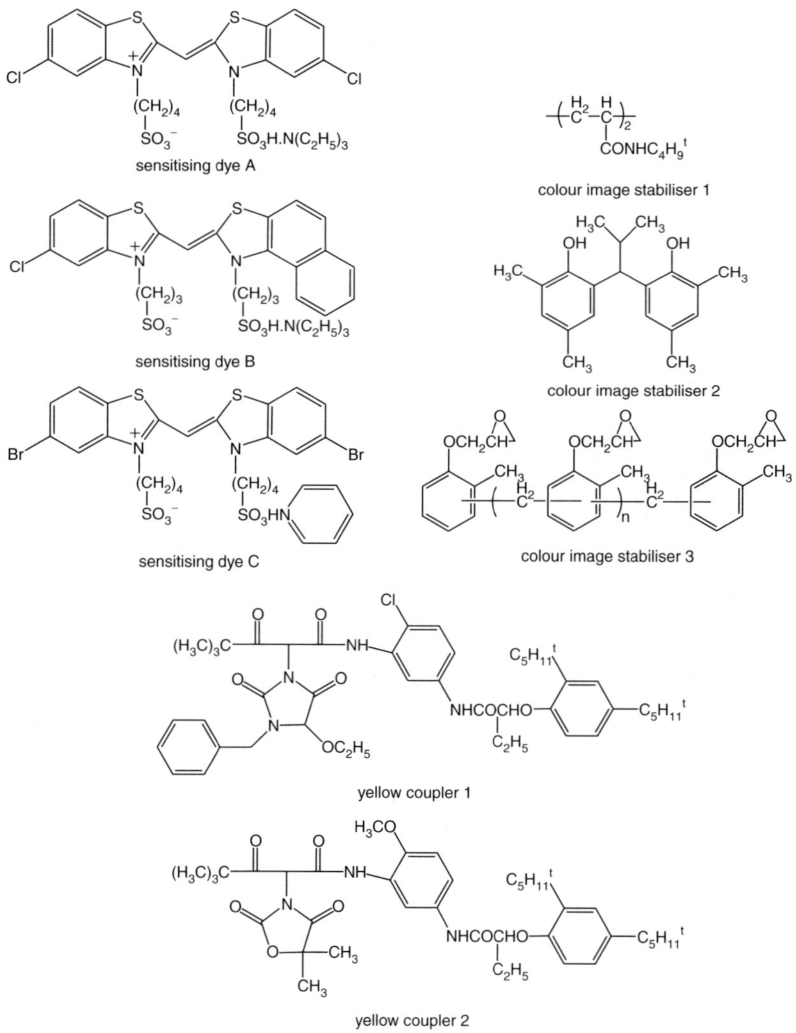

Figure 8 *Components of the blue sensitive layer*

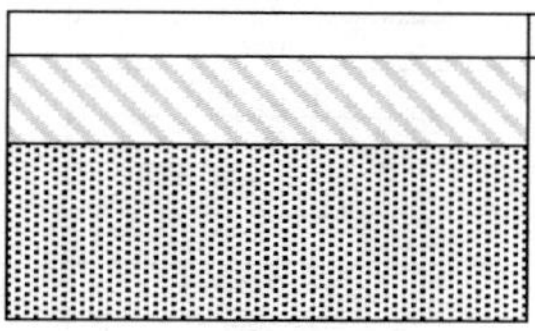

Component	Laydown rate
Gelatin	0.95
Colour-mixing inhibitor	0.12
Stabiliser 2	0.007
Stabiliser 1	0.14
Stabiliser 3	0.006
Solvent 1	0.06
Solvent 2	0.22
1-(3-methylureidophenyl)-5 -mercaptotetrazole	0.2 mg/m^2
disodium catechol-3,5-disulphonate	6 mg/m^2

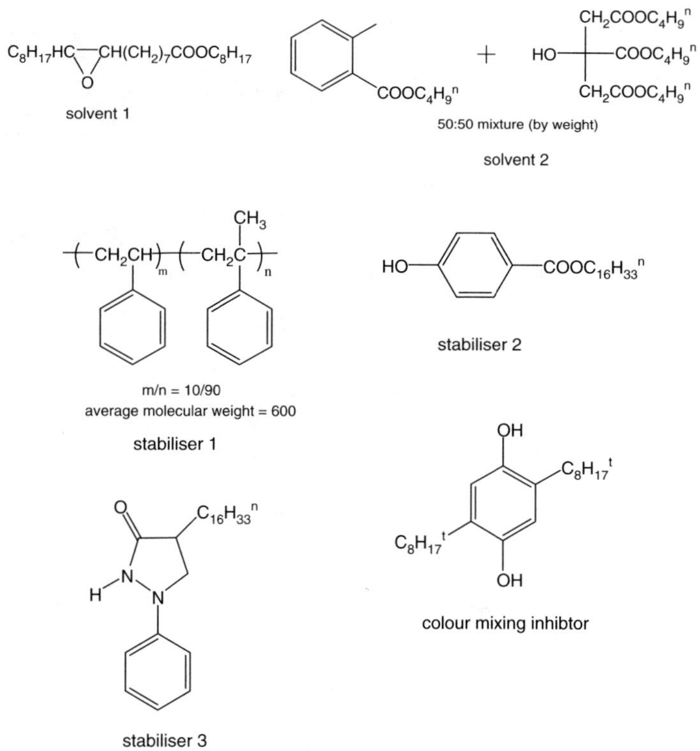

Figure 9 *Components of the first interlayer*

10.1.1 Blue Sensitive Emulsion Layer

The emulsion described for this layer was a silver chlorobromide emulsion having 0.33 M%, of a silver bromide locally contained in part of the grain surface whose substrate was made of silver chloride. The average silver halide grain size was 0.62 μm (Figure 8).

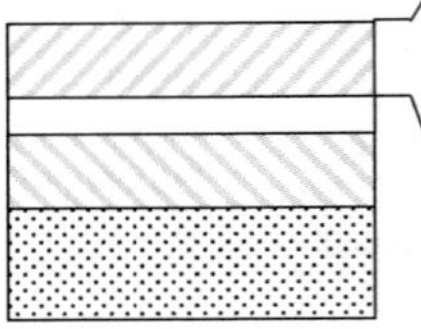

Component	Laydown rate
Gelatin	1.20
Magenta coupler 1	0.10
Magenta coupler 2	0.05
Ultraviolet absorbing agent 1	0.05
Ultraviolet absorbing agent 2	0.02
Ultraviolet absorbing agent 3	0.02
Ultraviolet absorbing agent 4	0.03
Colour-image stabiliser 2	0.005
Colour-image stabiliser 4	0.002
Colour-image stabiliser 1	0.08
Colour-image stabiliser 8	0.015
Colour-image stabiliser 9	0.03
Colour-image stabiliser 10	0.01
Colour-image stabiliser 11	0.0001
Colour-image stabiliser 13	0.004
Solvent 3	0.10
Solvent 4	0.19
Solvent Sol-5	0.17
1-(3-methylureidophenyl)-5-mercaptotetrazole	1.0×10^{-3} mol*
4-hydroxy-6-methyl-1,3,3a,7-tetrazaindene	2×10^{-4} mol*

* per mole of the silver halide

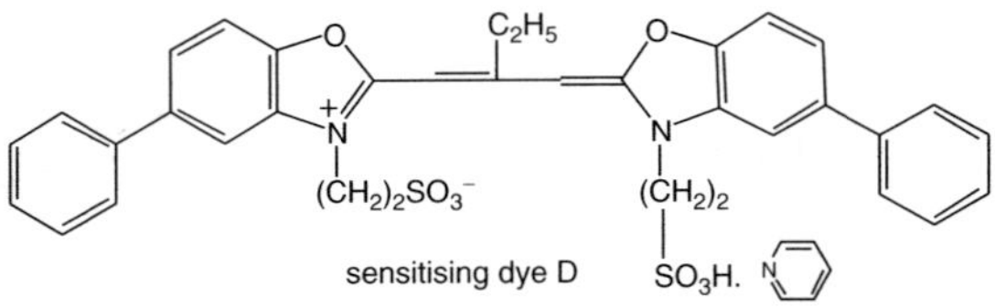

sensitising dye D

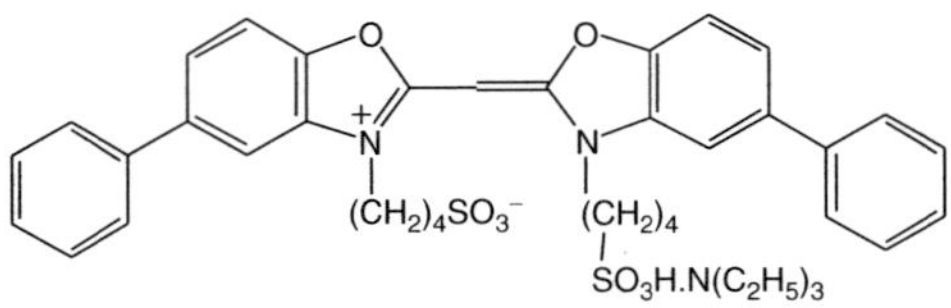

sensitising dye E

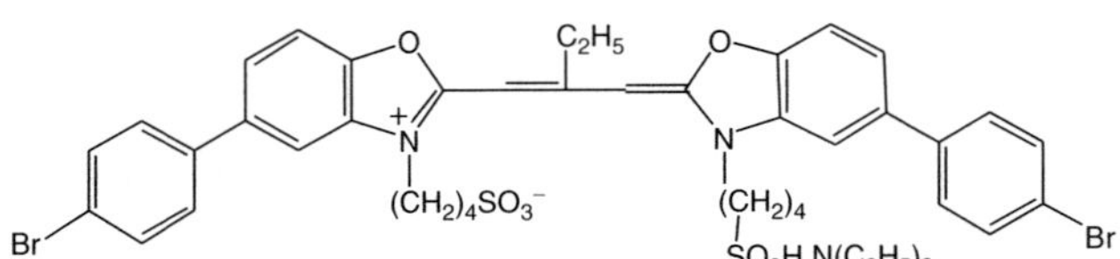

sensitising dye F

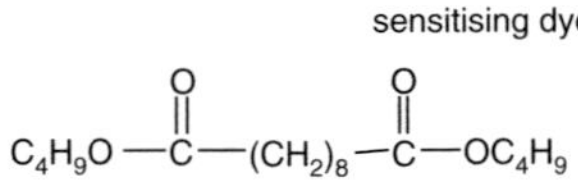

solvent 3

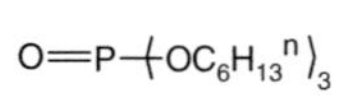

solvent 4

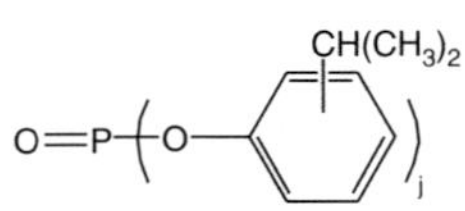

solvent 5

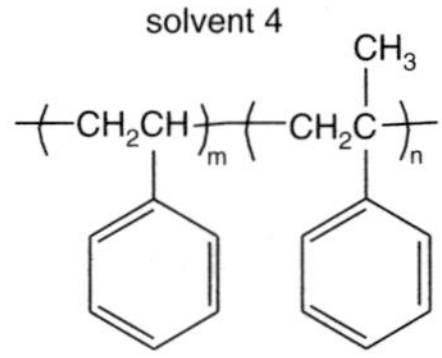

m/n = 10/90
average molecular weight = 600

stabiliser 1

Figure 10 *Components of the green sensitive layer*

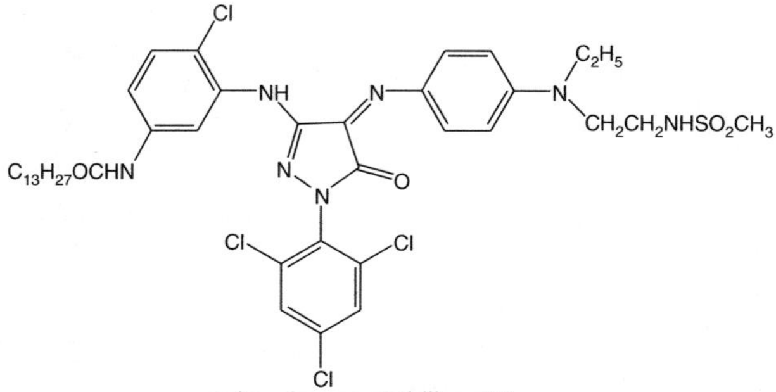

colour image stabiliser 11

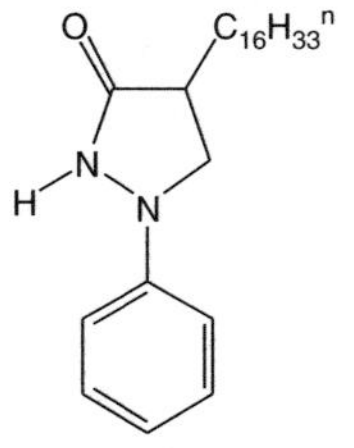

colour image stabiliser 13

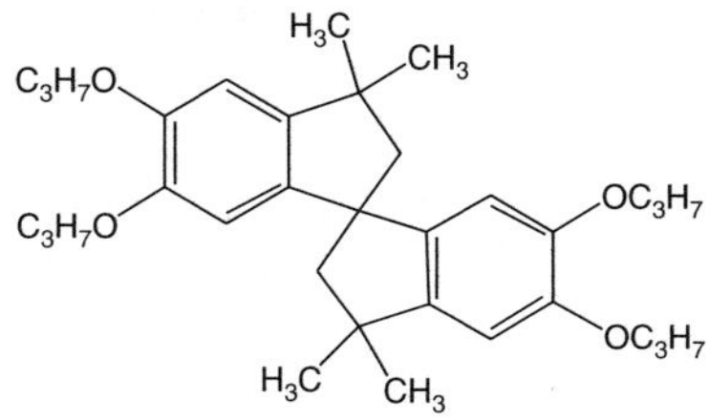

colour image stabiliser 8

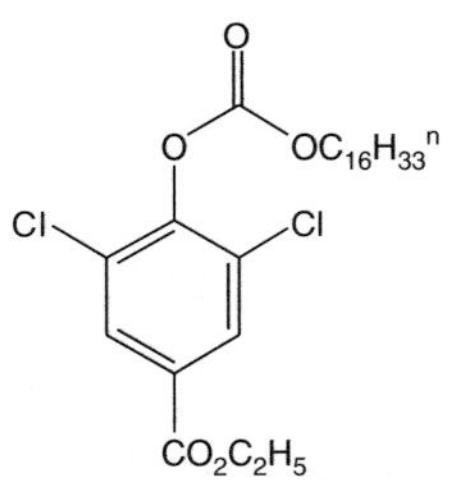

colour image stabiliser 9

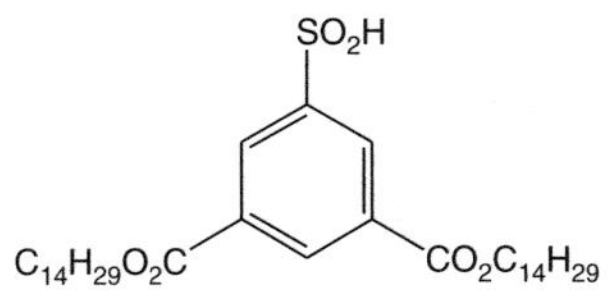

colour image stabiliser 10

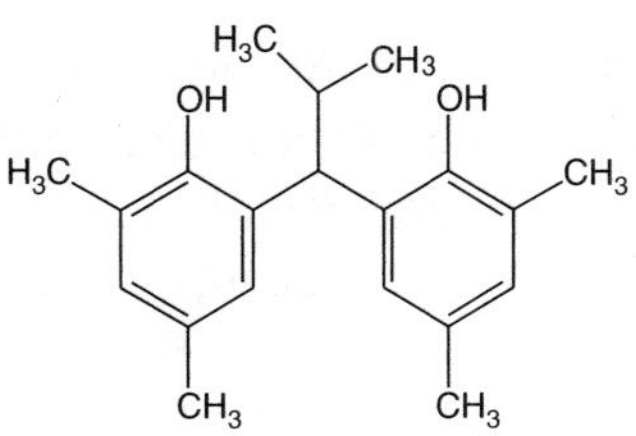

colour image stabiliser 2

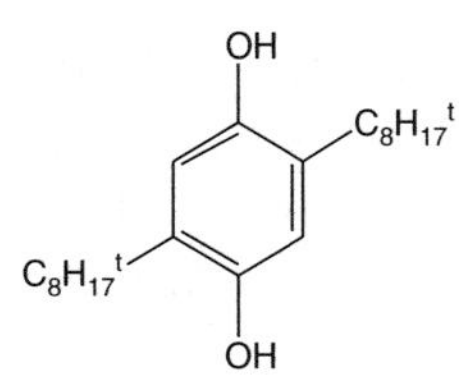

colour image stabiliser 4

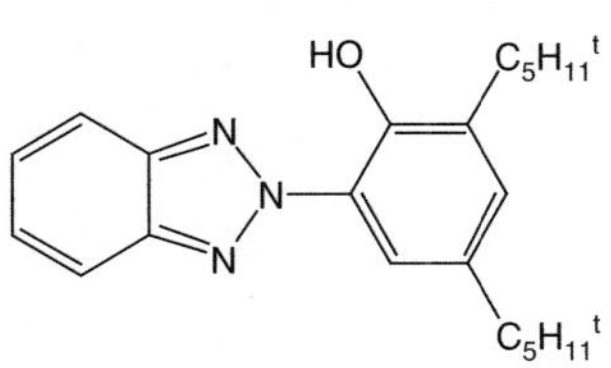

ultraviolet absorbing agent 1

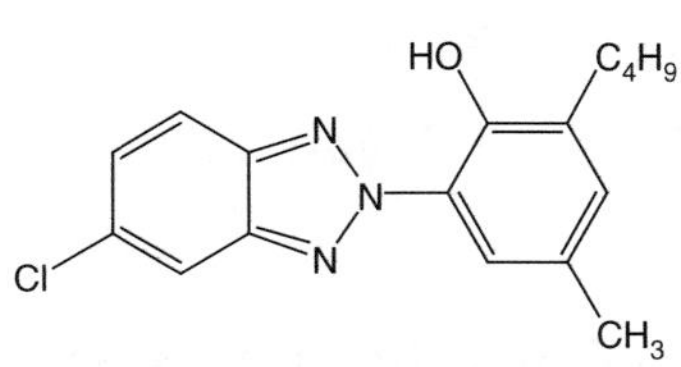

ultraviolet absorbing agent 2

Figure 10 (*Continued*)

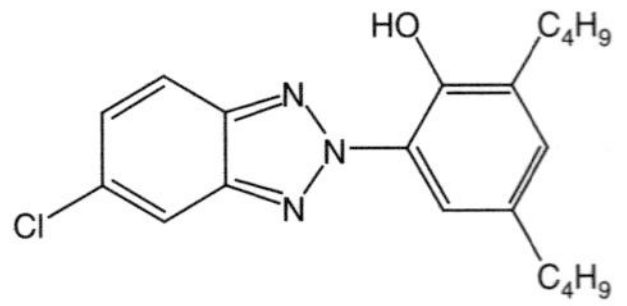

ultraviolet absorbing agent 3

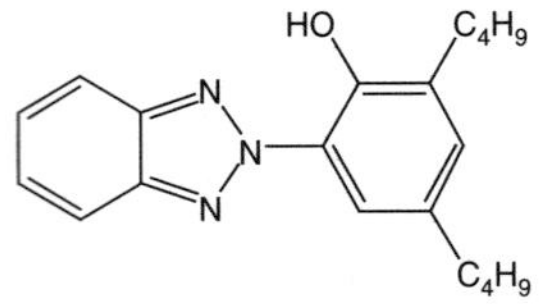

ultraviolet absorbing agent 4

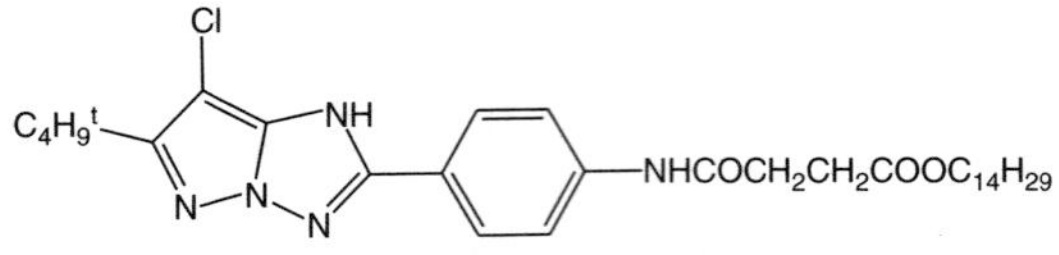

magenta coupler 1

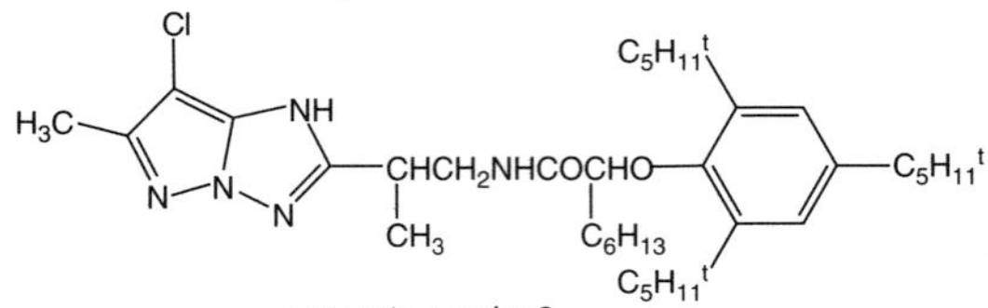

magenta coupler 2

Figure 10 (*Continued*)

10.1.2 Interlayer

For interlayer, see Figure 9.

10.1.3 Green Sensitive Emulsion Layer

A mixture of two silver chlorobromide cubic emulsions (0.42 μm for the larger and 0.33 μm for the smaller). Each emulsion had 0.69 mol% and 0.81 mol%, respectively, of a silver bromide locally contained in part of the grain surface, whose substrate was made up of silver chloride (Figure 10).

10.1.4 Interlayer

For interlayer, see Figure 11.

10.1.5 Red Sensitive Emulsion Layer

A mixture of two chlorobromide silver halide cubic emulsions, the larger with an average edge length of 0.41 μm, and the smaller an average edge length of 0.34 μm (Figure 12).

Interlayer

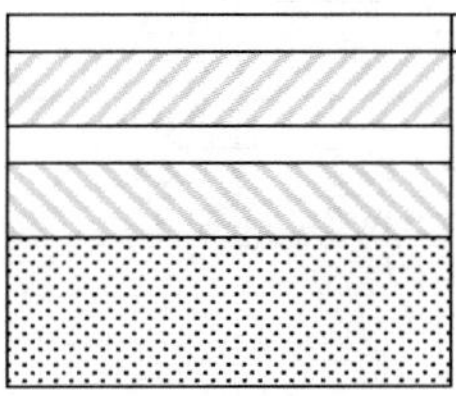

Component	Laydown rate
Gelatin	0.71
Colour-mixing inhibitor	0.09
Stabiliser 2	0.005
Stabiliser 1	0.10
Stabiliser 3	0.004
Solvent 1	0.04
Solvent 2	0.16
1-(3-methylureidophenyl)-5-mercaptotetrazole	0.2 mg/m^2
disodium catechol-3,5-disulphonate	6 mg/m^2

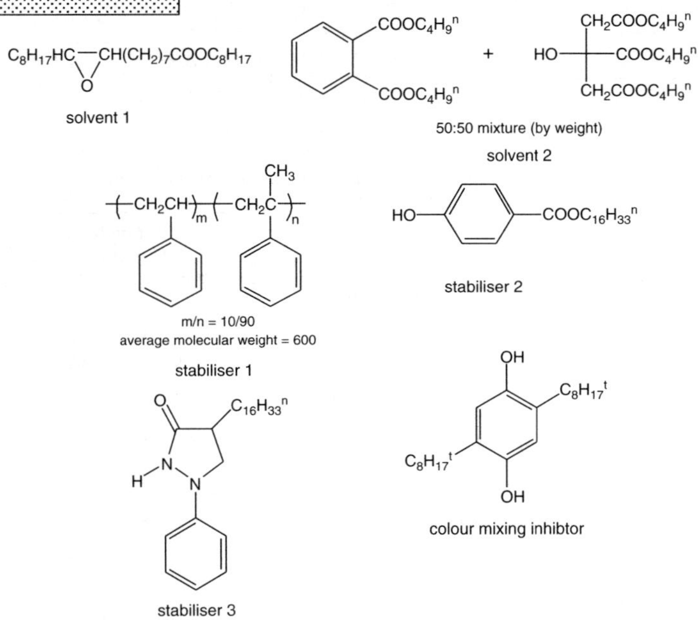

Figure 11 *Components of the second interlayer*

10.1.6 Ultraviolet Filter Layer

For ultraviolet filter layer, see Figure 13.

10.1.7 Protective Overcoat with Matte Beads

For protective overcoat with matte beads, see Figure 14.

10.2 Common Components

Sodium 1-oxy-3,5-dichloro-s-triazine was used as gelatin hardener. Figure 15 details those components and their laydown rates (in mg m^{-2}) that are common to more than two layers.

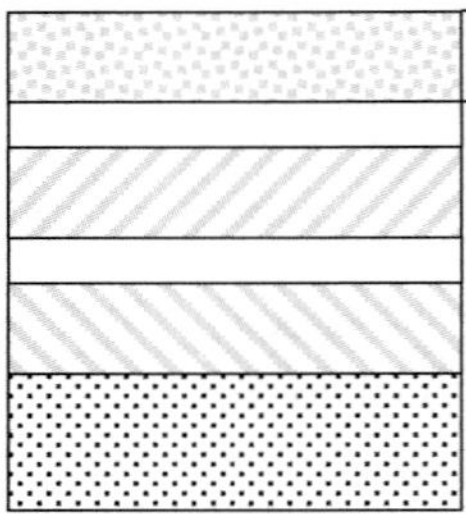

Component	Laydown rate
Gelatin 1.00	
Cyan coupler (ExC-1)	0.05
Cyan coupler (ExC-2)	0.18
Cyan coupler (ExC-3)	0.024
Ultraviolet absorbing agent (UV 1)	0.04
Ultraviolet absorbing agent (UV-3)	0.01
Ultraviolet absorbing agent (UV-4)	0.01
Colour-image stabiliser (Cpd-1)	0.23
Colour-image stabiliser (Cpd-9)	0.01
Colour-image stabiliser (Cpd-12)	0.01
Colour-image stabiliser (Cpd-13)	0.006
Solvent (Solv-6)	0.23
Compound 1	2.6×10^{-3} mole*
1-(3-methylureidophenyl)-5-mercaptotetrazole	5.9×10^{-4} mol*
copolymer of methacrylic acid and butyl acrylate^	0.05

* per mole of the silver halide
^1:1 in weight ratio; average molecular weight, 200,000 to 400,000

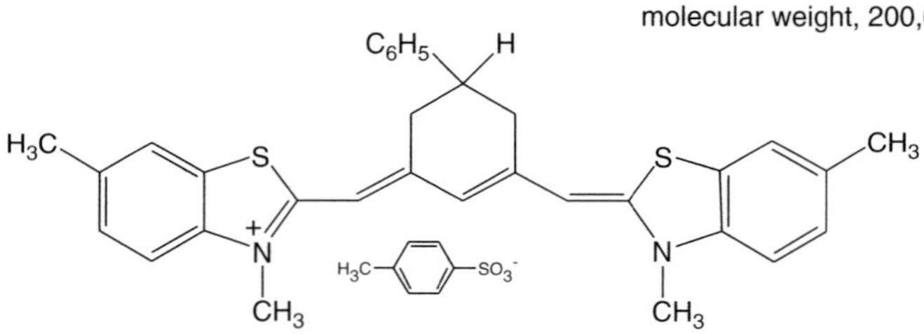

red sensitising dye

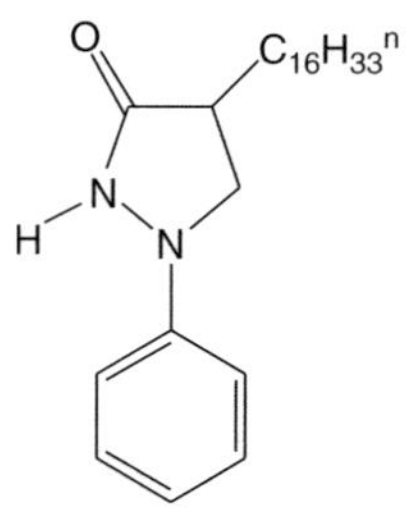

colour image stabiliser 3

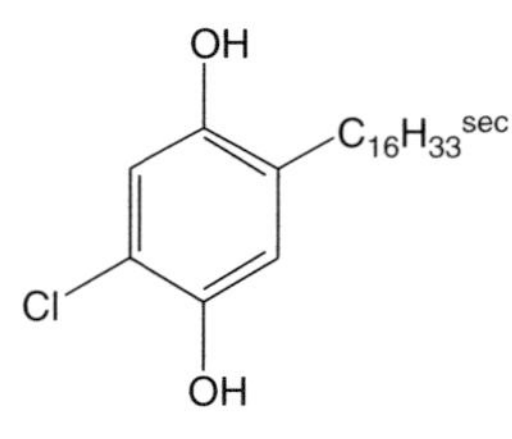

colour image stabiliser 12

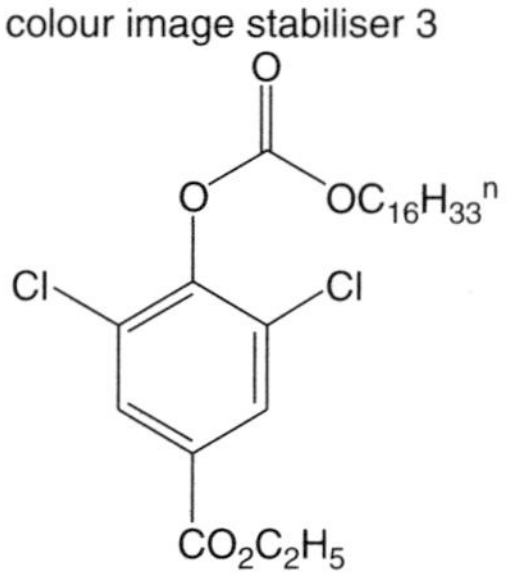

colour image stabiliser 9

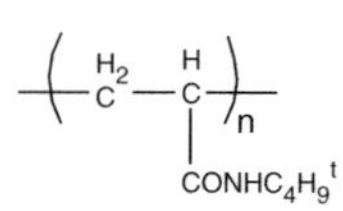

colour image stabiliser 1

Figure 12 *Components of the red sensitive layer*

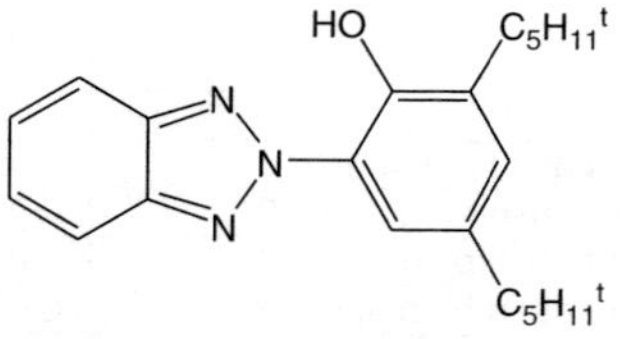

ultraviolet absorbing agent 1

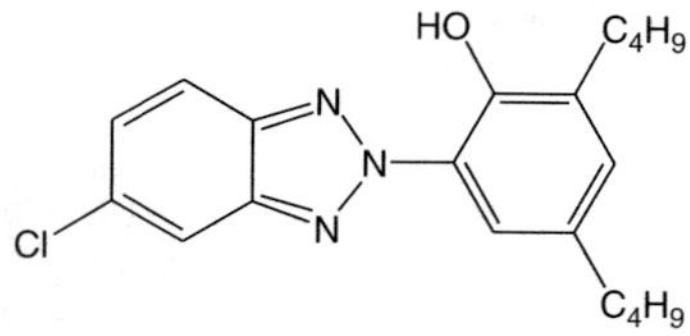

ultraviolet absorbing agent 3

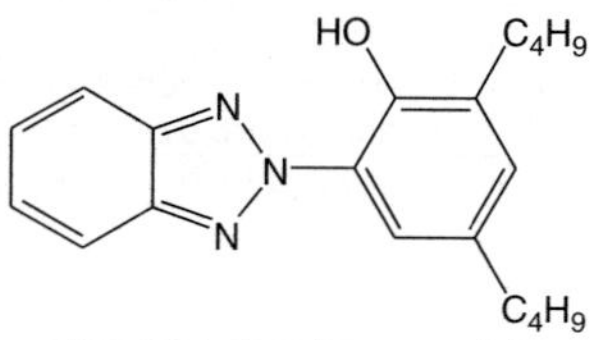

ultraviolet absorbing agent 4

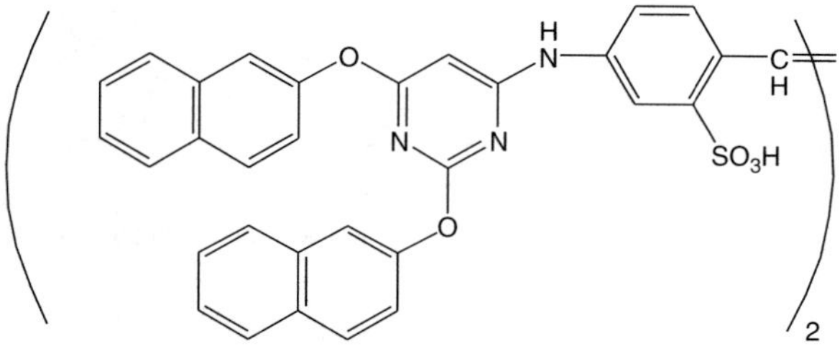

compound I

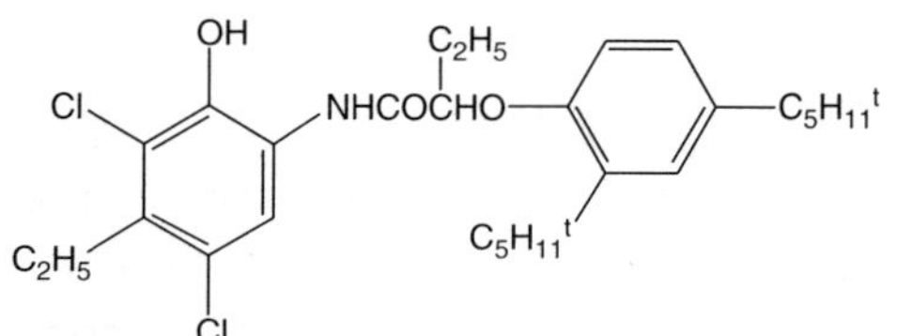

cyan coupler 1

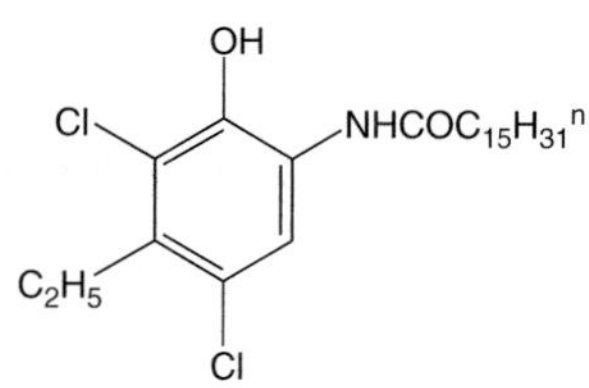

cyan coupler 2

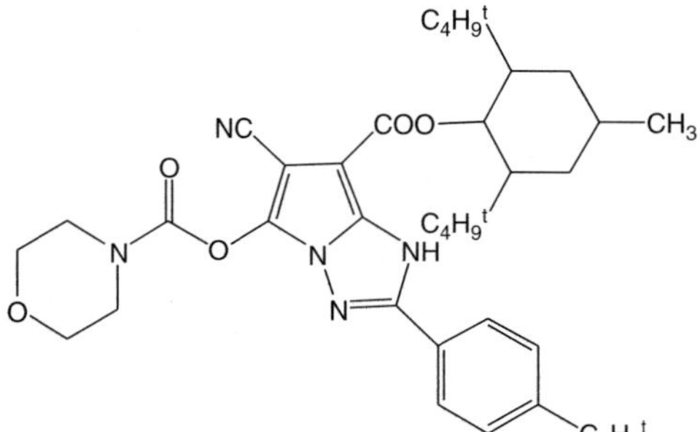

cyan coupler 3

Figure 12 (*Continued*)

Ultraviolet Filter Layer

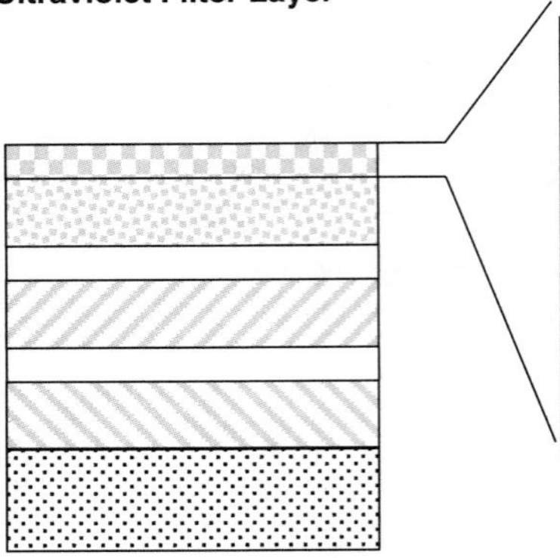

Component	Laydown rate
Gelatin	0.34
Ultraviolet absorbing agent 1	0.08
Ultraviolet absorbing agent 2	0.03
Ultraviolet absorbing agent 3	0.03
Ultraviolet absorbing agent 4	0.02
Ultraviolet absorbing agent 5	0.01
Ultraviolet absorbing agent 6	0.03
Solvent 7	0.10
1-(3-methylureidophenyl)-5-mercaptotetrazole	0.6 mg/m^2
disodium catechol-3,5-disulphonate	18 mg/m^2

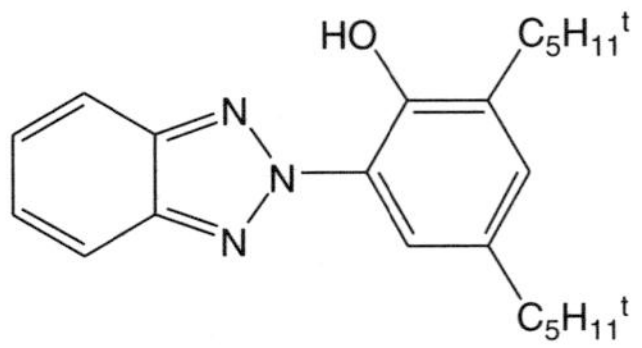

ultraviolet absorbing agent 1

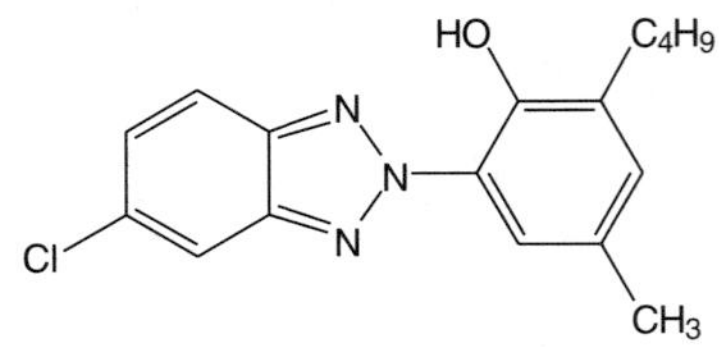

ultraviolet absorbing agent 2

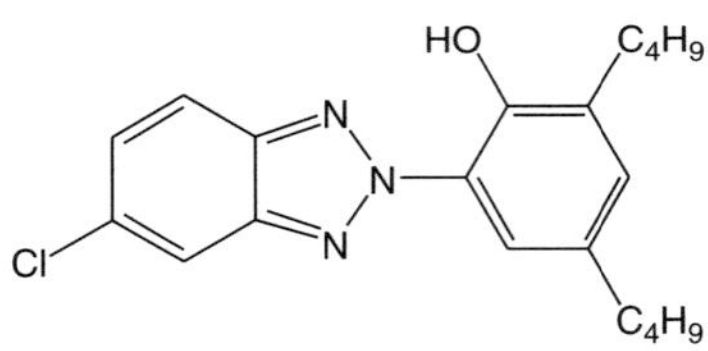

ultraviolet absorbing agent 3

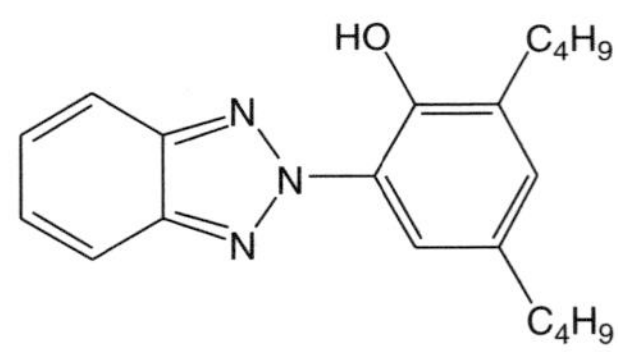

ultraviolet absorbing agent 4

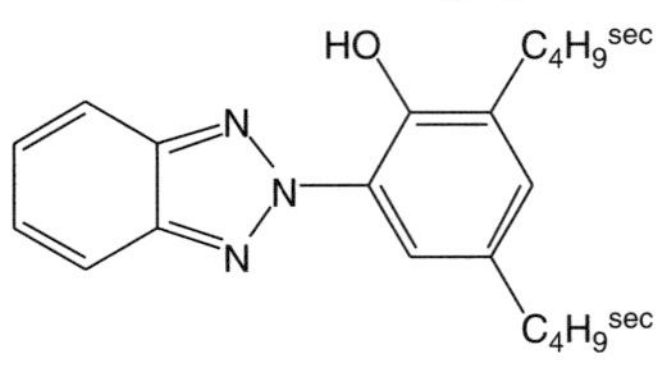

ultraviolet absorbing agent 5

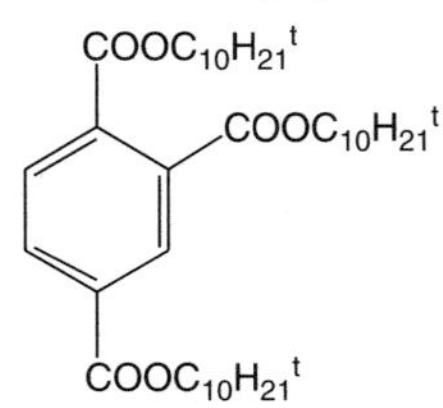

solvent 7

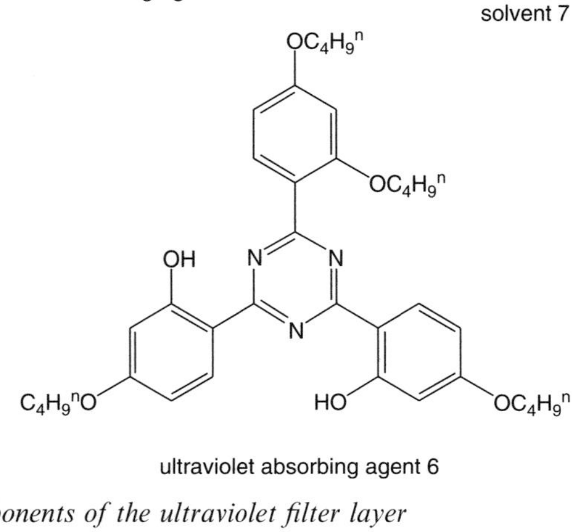

ultraviolet absorbing agent 6

Figure 13 *Components of the ultraviolet filter layer*

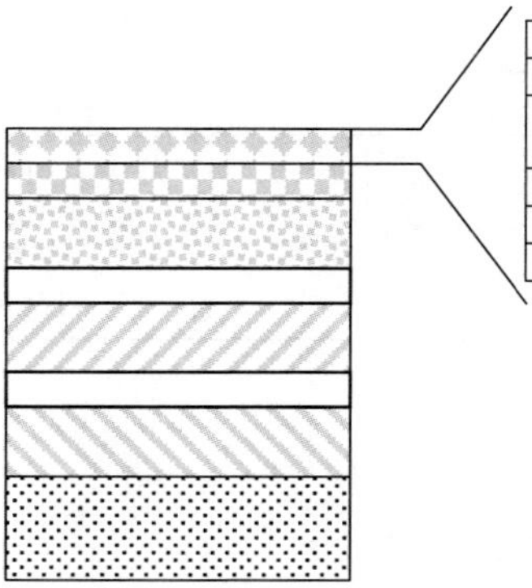

Component	Laydown rate
Gelatin	1.00
Acryl-modified copolymer of polyvinyl alcohol (modification degree: 17%)	0.04
Liquid paraffin	0.02
Surface-active agent (Cpd-14)	0.01
Surface-active agent (Cpd-15)	0.01

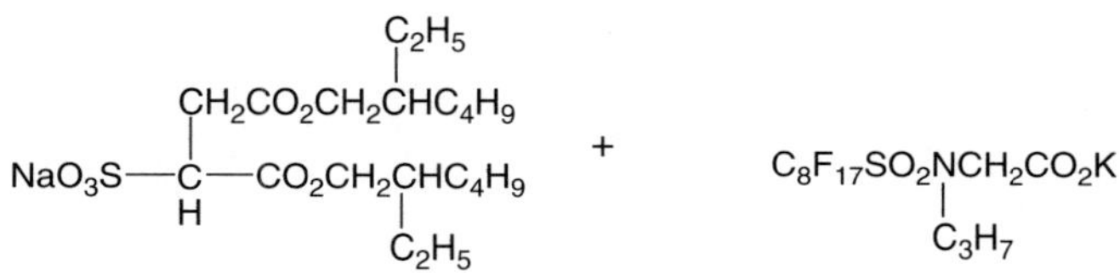

A mixture of 7:3 by weight

surfactant 14

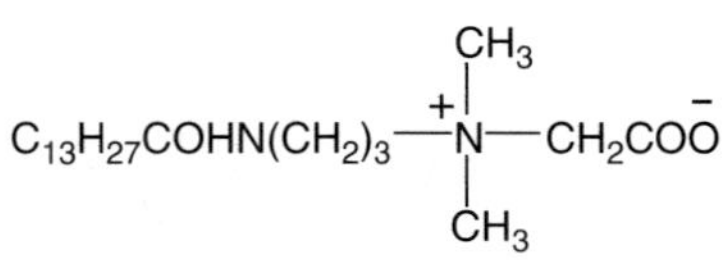

surfactant 15

Figure 14 *Components of the protective overcoat layer*

Compounds A–E are added for irradiation protection [Figure 15(b)].

The four antiseptic additives, labelled Ab 1 to 4 are added to each layer so that the total amounts are 15.0 mg m^{-2}, 60.0 mg m^{-2}, 5.0 mg m^{-2} and 10.0 mg m^{-2}, respectively (Figure 16).

Ab 4 is a mixture of four components a, b, c and d and is used in the ratio 1:1:1:1, see Table 1.

Table 1 *Components of Ab 4*

	R_1	R_2
a	$-CH_3$	$-NHCH_3$
b	$-CH_3$	$-NH_2$
c	$-H$	$-NH_2$
d	$-H$	$-NHCH_3$

(a)

Layer	A	B	C	D	E
Red sensitive	2	5	1	20	10
Green sensitive	2	5	1	20	10
Blue sensitive	2	5	1	20	10

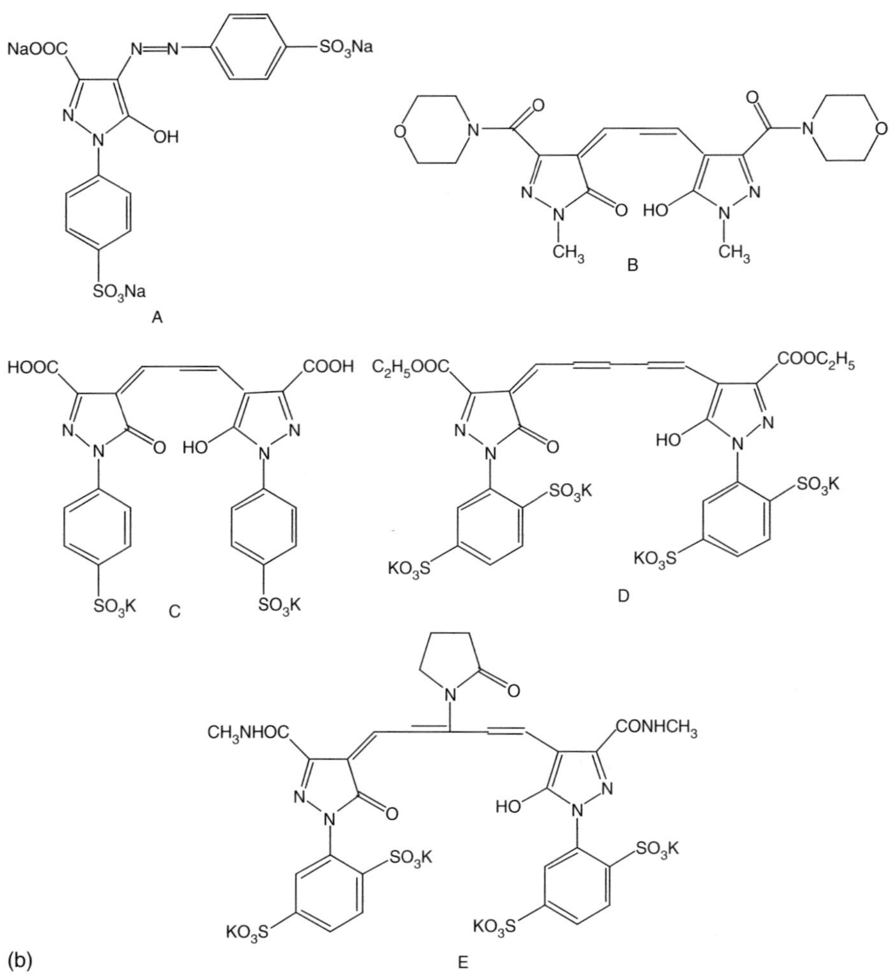

(b)

Figure 15 *Common components*

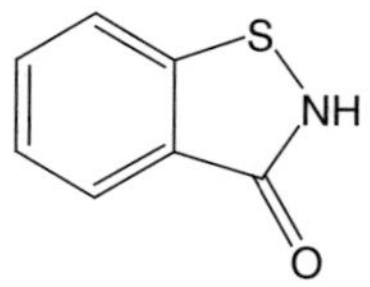

Ab 1

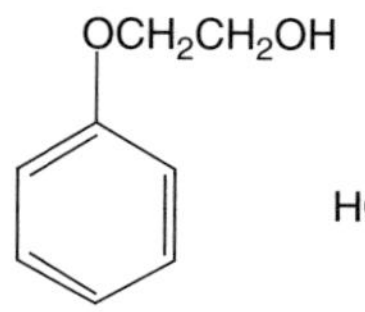

Ab 2

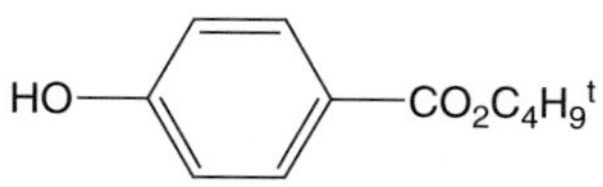

Ab 3

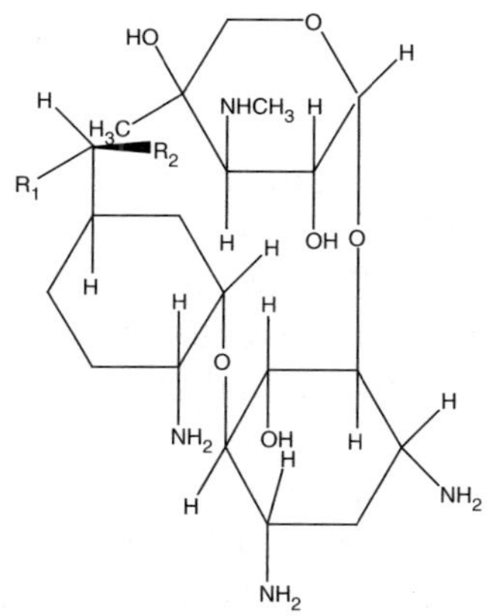

Ab 4

Figure 16 *Antiseptic additives*

References

1. R.N. Govers and I. De Vries, US 2003/0113675, Fuji Photo Film B.V.
2. J.B. Bouwstra and I. De Vries, EP 1,130,460, Fuji Photo Film B.V.
3. R. Edwards, US 4,526,919, Eastman Kodak Co.
4. J.C. Diaz and R.A. Mears, US 3,652,725, Eastman Kodak Co.
5. K.A. Ridley, US 4,135,477, Ciba-Geigy AG.
6. A. Wiley Clayton, US 4,075,976.
7. J.F. Greiller, US 3,968,772, Eastman Kodak Co.
8. J.E. Conroy and K.J. Ruschak, US 5,358,569, Eastman Kodak Co.
9. M. Slagt, A. Kase and P. van Asten, WO 2004081661, Fuji Photo Film B.V.

CHAPTER 11

Kodachrome Films

Kodachrome was the first colour product aimed at the amateur market by any photographic manufacturer. It was launched on the 15th April 1935 as a 16 mm cine film and required 28 separate processing steps when first introduced. It was invented by Leopold Godowsky and Leopold Mannes, who were professional musicians and amateur inventors.

They worked on their invention from the early 1920s, but signed up to work for Eastman Kodak Co. on 31st October 1930 until 1939, when they both returned to their work as musicians. From the very outset they determined to include the couplers in the processing steps, see for example their joint patents.[1–3] In his speech to the 30th annual conference of the Society of Photographic Scientists and Engineers, delivered on the 6th May 1977, W.T. Hanson Jr, then director of the Eastman Kodak Research Laboratories, commented of the Kodachrome process[4]

> *... The 16 mm Kodachrome film introduced in 1935 consisted of six layers on a single support. The top emulsion layer was blue sensitive; to avoid exposing incorrectly the emulsions coated beneath, it contained a yellow dye to absorb all blue light that entered it. Below the blue sensitive layer was a layer of clear gelatin. This was important. Below that was a blue-and green sensitive layer, another gelatin interlayer, and finally a blue- and red sensitive layer. Thus, three separation latent images were made when the film was exposed, blue in the top layer, green in the middle, and red in the bottom. The processing followed from the film structure. First the silver negative images in all three sensitised layers were developed simultaneously, and the yellow filter dye was removed from the blue sensitive top layer. The developed silver was then bleached and removed, and the remaining positive images in sensitive silver halide were exposed to light. All three layers were then developed in a cyan dye-forming color developer. With the cyan image completed, the film was dried, wound onto spools, and removed from the processor. In another processor, the developed silver and cyan dye in the top two layers were bleached; the silver was rehalogenized and the dye*

destroyed. This was the key to the process and was known as controlled diffusion bleach. By timing it right and starting with dry film, Mannes and Godowsky were able to bleach the top layers completely without affecting the bottom layer. The gelatin interlayer that provided a margin of safety was one ten-thousandth of an inch thick (2.5 micrometers). After bleaching, the film was exposed and developed again. Silver images were developed in the top two layers, and the oxidised developer reacted with a coupler to form magenta dye. Again, the film was dried, spooled, and fed into another machine that repeated the process of bleaching, rehalogenising, exposing and developing to form yellow dye. Finally, all of the developed silver was removed. In all, 28 steps were involved.

An important modification in the process was made in 1938; the drying and controlled diffusion bleach were replaced with selective re-exposure for the color-development steps. The total number of steps was reduced from 28 to 18 . . .

The modern schematic diagram for images produced through this colour reversal process appears as Figure 15 in Chapter 1. The modern processing steps involve

- first development
- dye development
- bleach and fix.

Figure 1 shows the sensitometry for a typical Kodachrome low-speed film. The red, green and blue curves have been plotted as grey curves, where the unmarked curve represents the green sensitometry.

This sensitometry is obtained using yellow magenta and cyan dyes of slightly different spectral responses, compared with the colour negative film and of colour paper. Figure 2 details the dye curves.

Modern Kodachrome films have the film structure as outlined in Figure 3, where the protective layers have been omitted for convenience.

The following tables record the processing solutions, which are known as K-14M. Some of the couplers are present as solutions, Figure 4.

The chemical constituents of processing solutions are in the public domain through the published materials safety data sheets (MSDSs), which accompany each shipment of liquids. These MSDS files contain data concerning the chemical components and approximate concentrations, the toxicity and spillage clean-up procedures. Some of the K-14M MSDSs were relatively easy to obtain from the Eastman Kodak Co. website,[5] the rest of the MSDSs which appear below were obtained from.[6] Still further data may be obtained from the University of Hawaii website.[7]

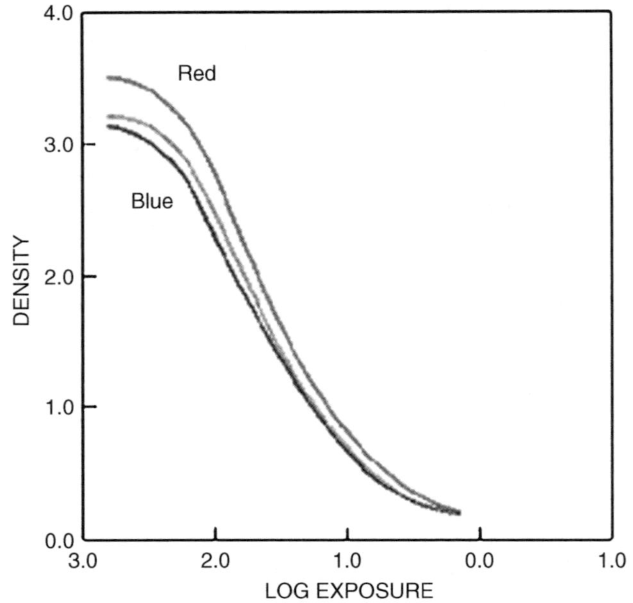

Figure 1 *Density vs log exposure curves for Kodachrome film*

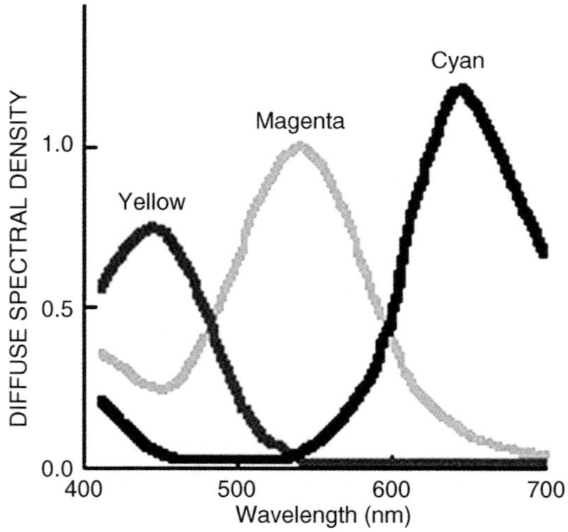

Figure 2 *Spectral responses for Kodachrome film*

The use of Remjet as a mean of protecting a film from the effects of halated or back-reflected light was discussed in Chapter 9 Figure 12. The first of the processing solutions softens the Remjet and then removes the Remjet backing layer, using a buffered solution of the pentasodium salt

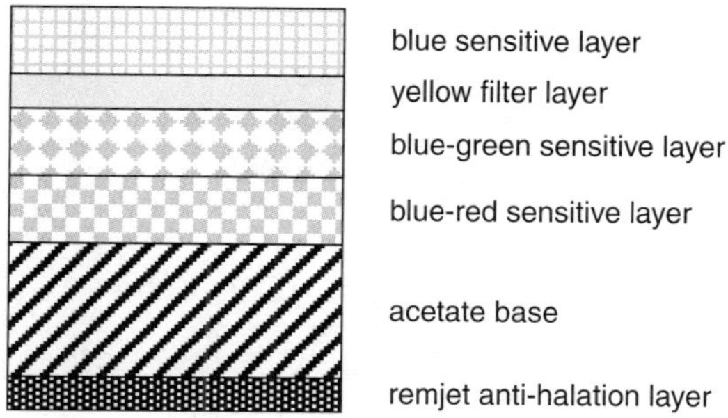

Figure 3 *Layer order in Kodachrome films*

Solution / Step	Time (min:sec)
Remjet removal	0:10
Rinse	0:15
First developer	2:00
Wash	1:00
Red re-exposure printer	
Cyan developer	2:00
Wash	2:00
Blue re-exposure printer	
Yellow developer	3:00
Wash	2:00
Magenta developer	4:00
Wash	2:00
Conditioner	1:00
Bleach	5:00
Fixer	3:00
Wash	2:00
Final rinse	1:00
Dryer	6:00

Figure 4 *Processing steps for the K-14M process*

of pentetic acid. Thereafter the series of processing steps detailed in Figure 4 results in a colour positive image.

Figure 5 shows a typical exposure through a filter, half of which is black, the other half of which is clear. All three layers will be exposed with the white light part of the image.

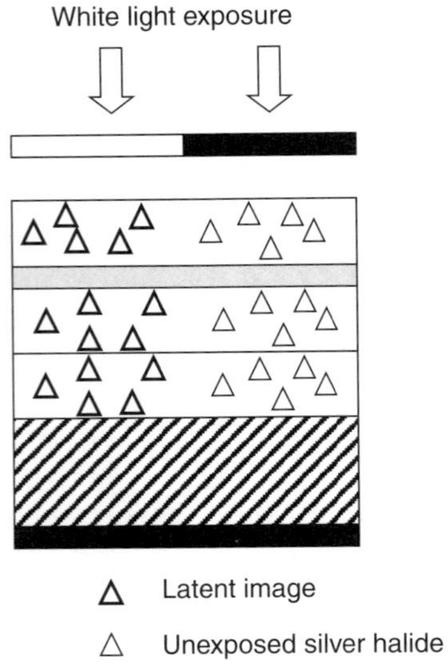

Figure 5 *White light exposure of a Kodachrome film*

11.1 First Developer Solution

In the first developer solution, the exposed silver halide grains (latent images) are reduced to metallic silver, by the action of Phenidone† and hydroquinone developers

$$\text{Exposed } Ag^{+} + \text{Developer} \longrightarrow Ag^{0} + \text{oxidized Developer} + \text{halide ions}$$

The resulting silver grains form three superimposed negative images of the original scene. The remaining unexposed and undeveloped silver halide in the three emulsion layers constitute the positive, reversal, image that is later converted to a full-colour image in the colour-development phases of the process, Figure 6.

The chemical composition of the developer solution is provided in Figure 7.[6]

11.2 Red Re-Exposure Printing Step

A red-exposure printing step through the film base completely exposes all of the remaining silver halide in the bottom emulsion layer, in order to ensure that the silver halide develops completely in the cyan developer

† Phenidone is an Ilford trade name and is one of the black and white developers mentioned in Chapter 4, Figure 1 as Phenidone Z.

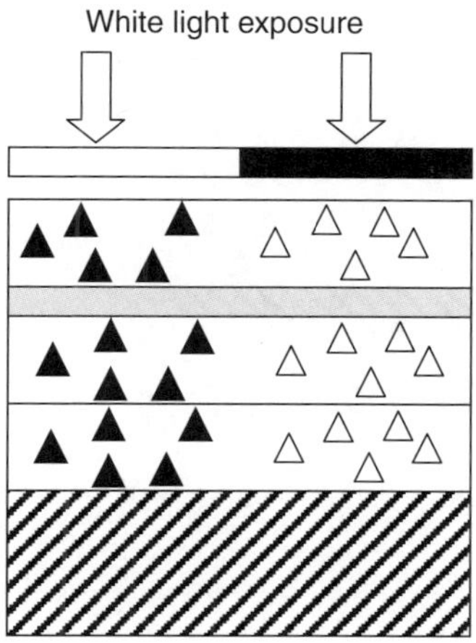

Figure 6 *The effect of first developer solution*

Weight %	Component
90-95%	Water
1-5%	Sodium carbonate
<1%	Sodium bicarbonate
<1%	Hydroquinone

Figure 7 *The chemical composition of the developer*

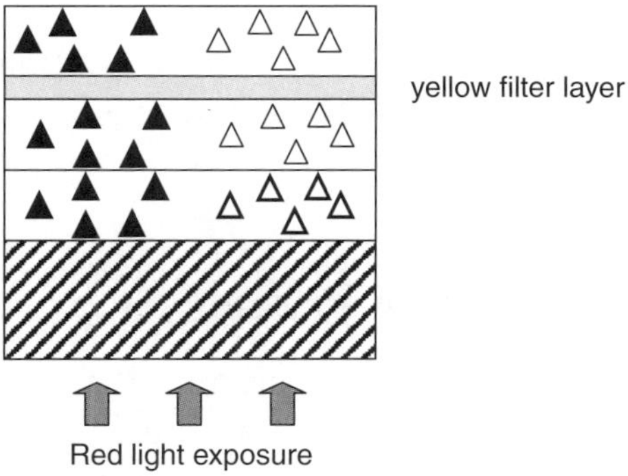

Figure 8 *Film exposure through the base*

solution, Figure 8. Care must be taken so that the red light does not expose any silver halide in the other layers. Accordingly, this red exposure uses a red glass filter in the light beam, which has been specifically designed to avoid as much unwanted exposure as possible. This selective exposure however, does result in some unwanted green exposure, which is compensated for by control of the red printing intensity.

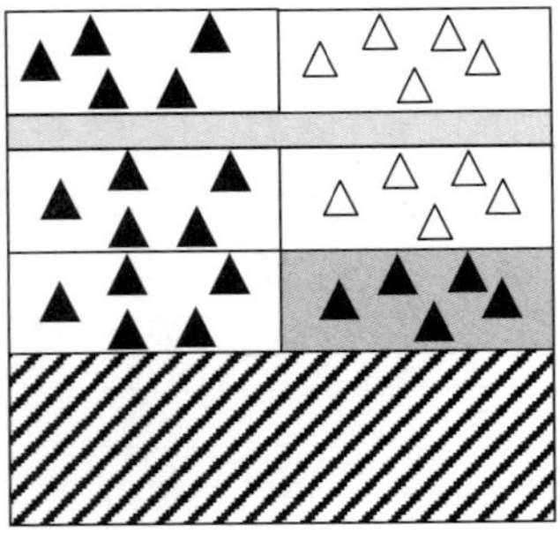

Figure 9 *Cyan dye formation*

Weight %	Component
95-100%	water
1-5%	sodium sulphate
1-5%	sodium sulphite
< 1%	sodium hydroxide
< 1%	4-(N-ethyl-N-2-hydroxyethyl)-2-methylphenylenediamine sulphate

Figure 10 *Chemical components of the cyan developer solution*

11.3 Cyan Developer Solution

The cyan developer solution creates a positive silver image and oxidised colour developer, which reacts with the cyan coupler to form cyan dye, Figure 9.

The chemical components of the working solution, Figure 10,[6] show that the colour developer is CD4. The reaction of oxidised developer with colour couplers was discussed in Chapter 6, Figures 4 and 5.

11.4 Blue Re-Exposure Printing Step

During this step, a blue light exposure through a blue glass filter from the front surface of the film creates a latent image in the yellow, uppermost layer. A yellow filter layer prevents the blue light form exposing the blue-green sensitive layer, Figure 11.

11.5 Yellow Developer Solution

A positive silver image is formed creating oxidised colour developer, which in turn reacts with the yellow coupler to form a yellow dye. This reaction must be complete in order to prevent contamination when the magenta layer is processed, Figure 12.

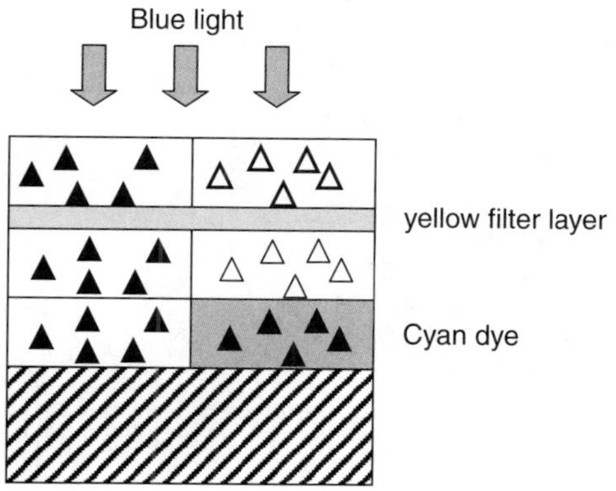

Figure 11 *Blue light exposure*

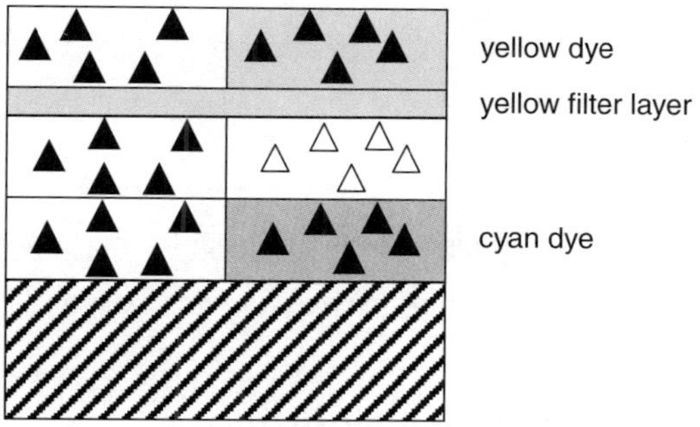

Figure 12 *Yellow dye formation*

Weight %	Component
	Bottle 1
100	2- (p-carboxyphenoxy) -2-pivaloyl-2', 4'-dichloroacetanilide
	Bottle 2
85-90	Water
10-15	2-methylpentane-2,4-diol
	Bottle 3
80-85	Water
15-20	Sodium hydroxide

Figure 13 *Yellow developer components*

The yellow developer components are supplied in three bottles, Figure 13.

It is interesting to note that the yellow coupler is similar to those used in both colour paper and colour film, mentioned in earlier chapters.

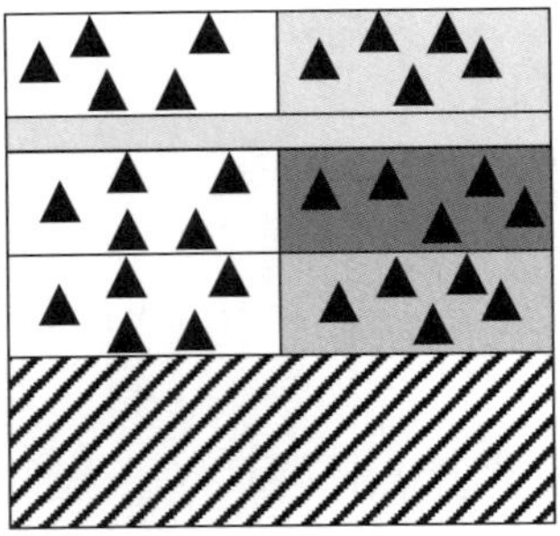

Figure 14 *At this stage all layers contain appropriate dyes*

Weight %	Component
90-95	water
1-5	sodium carbonate
1-5	sodium sulfate
< 1	sodium bicarbonate
< 1	sodium hydroxide
< 1	4-(N-ethyl-N-2-methanesulfonylaminoethyl)-2-methylphenylenediamine

Figure 15 *Working strength solution of the magenta developer*

11.6 Magenta Developer Solution

At this point the only unexposed silver halide crystals should be in the magenta layer. These are chemically exposed using a nucleating agent, and the magenta dye formed follows the same reactions as above, *i.e.* silver halide reducing to silver, as the colour developer is oxidised. This is followed by reaction of the oxidised colour developer with the magenta coupler, Figure 14.

This step is the most crucial as the final colour quality will be determined by the correct levels of magenta dye in the final image. Figure 15 shows the working strength solution for the magenta developer, which is CD3, unlike the cyan developer which is CD4, see Figure 10.[6]

11.7 Conditioner Solution

The film is run through a conditioner solution at this stage, the chemical components for which are provided in Figure 16.[5]

The EDTA in the conditioner solution will remove any iron or other metals that might otherwise compromise the bleach solution.

Weight %	Component
95-100	Water
1-5	Potassium sulphite
1	Ethylenediaminetetraacetic acid (EDTA)
0.1	1-thioglycerol

Figure 16 *Conditioner solution*

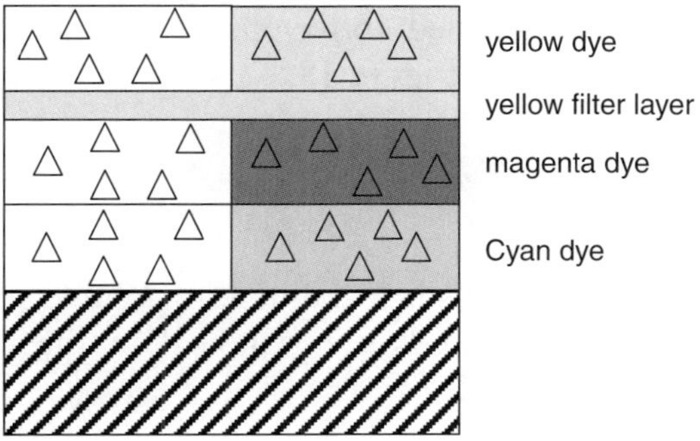

Figure 17 *Silver halide is present in all imaging layers*

Weight %	Component
55-60	Water
15-20	Ammonium ferric ethylenediaminetetraacetic acid
10-15	Ammonium bromide
1-5	Potassium nitrate
1-5	Hydrobromic acid
1-5	Acetic acid

Figure 18 *Working strength bleach solution*

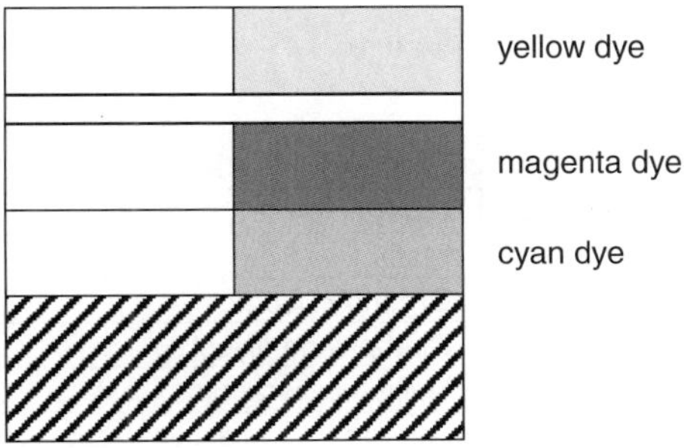

Figure 19 *Fully processed film*

11.8 Bleach

The bleach converts the metallic silver back to silver halide, which is later removed in the fixer. During bleaching, iron III is reduced to iron II. Iron II must be converted back to iron III by aeration so that satisfactory bleaching can continue. Indeed the bleach solution is aerated by bubbling air through it in order to preserve the life of the bleach solution. The yellow filter dye is also removed by the bleach, Figure 17.

The iron solution mentioned above contains ammonium ferric ethylenediaminetetraacetic acid, Figure 18.[5]

11.9 Fixer

The fixer converts all of the silver halide into soluble silver compounds. Most of the silver compounds are removed in the fixer and can be recovered, Figures 19 and 20.[6]

11.10 Final Rinse

The components for the final rinse are provided in Figure 21.[5]

The silicon component of the conditioning solution will allow the film to dry without leaving streak marks on the surface of the film. This is an important criterion as the film will be projected as either a still picture or a movie film (Kodachrome is marketed in both formats).

Weight %	Component
30-40	water
50-60	ammonium thiosulphate
1-5	sodium bisulphite
1-5	ammonium sulphite
1-5	sodium sulphite
<1	ethylenediamine-tetraacetic acid (EDTA)

Figure 20 *Working strength fixer solution*

Weight %	Component
95-99	Water
1-5	Organo-silicone
1-5	Dipropylene glycol
1	Nonionic surfactant
0.5	1,2-benzoisothiazolin-3-one

Figure 21 *Final rinse working solution*

References

1. L.D. Mannes and L. Godowsky, Jr., US 2,059,887, Eastman Kodak Co.
2. L.D. Mannes and L.Godowsky, Jr., US 2,091,713, Eastman Kodak Co.
3. L.D. Mannes and L. Godowsky Jr., US 2,039,730, Eastman Kodak Co.
4. Journey: 75 years of Kodak Research, Copyright Eastman Kodak Company, 1989.
5. www.kodak.com.
6. http://siri.org/msds/index.php.
7. http://kauai.hawaii.edu/msds/.

CHAPTER 12

Motion Picture Films

George Eastman, the founder of the Eastman Kodak Company, is often heralded as the father of popular photography and inventor of motion-picture film. He was born in 1854 and left school at age 14 to support his mother, Maria Kilbourn Eastman, and two elder sisters, Ellen and Kate. He founded the Dry Plate Company in 1881, which was subsequently re-incorporated as the Eastman Dry Plate and Film Company. Eastman patented and produced rollable film as an alternative to the glass negatives, which were the previous industry standard. He subsequently renamed his company Eastman Kodak Company in 1892. The use of the flexible support in film manufacture was the crucial invention in the production of motion picture film.

The requirement for special effects, multiple copies, and large projection format are just three of the needed features of this film family. These requirements resulted in a number of types of films as well as films of a number of photographic speeds within the family. Figure 1 outlines

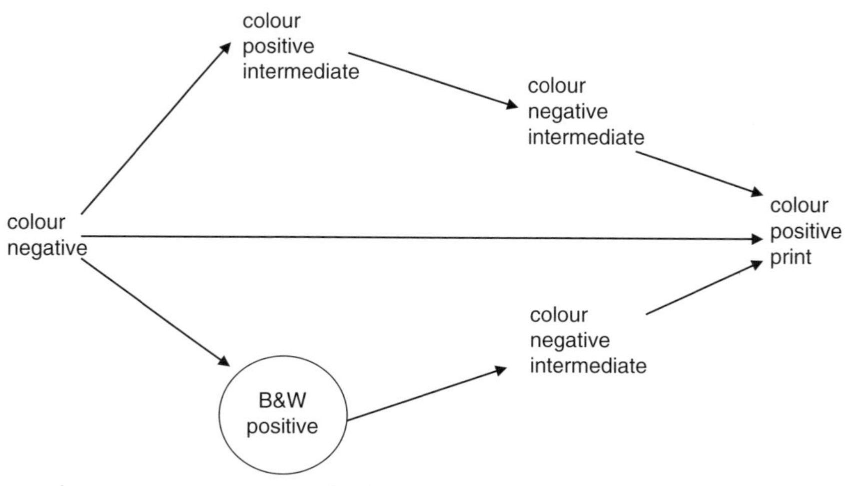

Figure 1 *The motion picture film 'family'*

some of the types of films that have been manufactured over the years demonstrating the complexity of this film family.

Figure 1 shows that there have been several possible routes from a colour negative to a colour positive print film. Some recent movies have been digitally shot and then distributed as colour positive print films. It is unlikely that mass market viewing will totally negate the use of films in the near future, partly because of the expense of converting every film viewing cinema over to a digital counterpart. As the cost of the digital hardware reduces, there may be more cinemas that convert but it is hard to imagine that some countries will be ready for the transition in the near future.

12.1 Colour Negative Film

Signal to noise is a large factor in the choice of a motion picture colour negative film. Ideally cinematographers would strive to use the slowest film stocks compatible with the available lighting. These films have a traditional film response of density plotted against log exposure, as outlined in Figure 2.

These types of films often have the conventional film layer order, format and chemistry,[1] see Figure 3.

12.2 Intermediate Film

The intermediate stages of producing a motion picture colour positive print film may involve either a black and white positive or a colour intermediate positive film, Figure 1. Figure 4 shows a typical colour

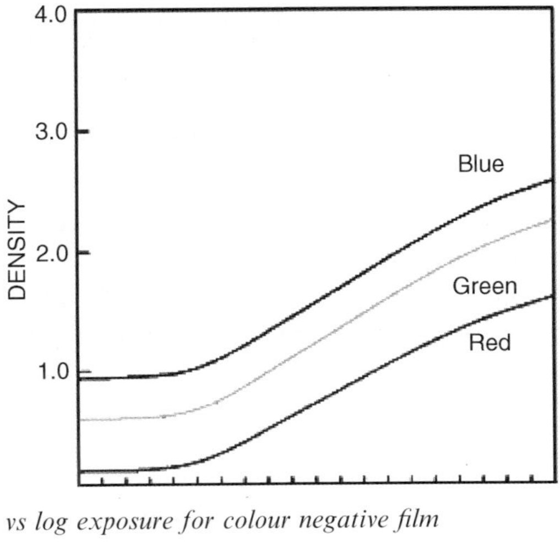

Figure 2 *Density vs log exposure for colour negative film*

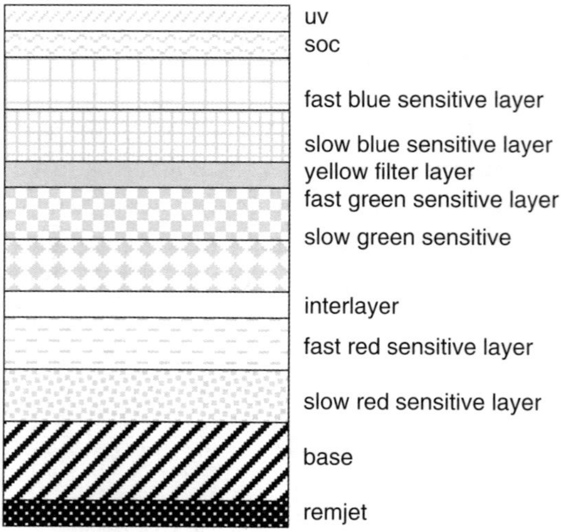

Figure 3 *Layer order for colour negative film*

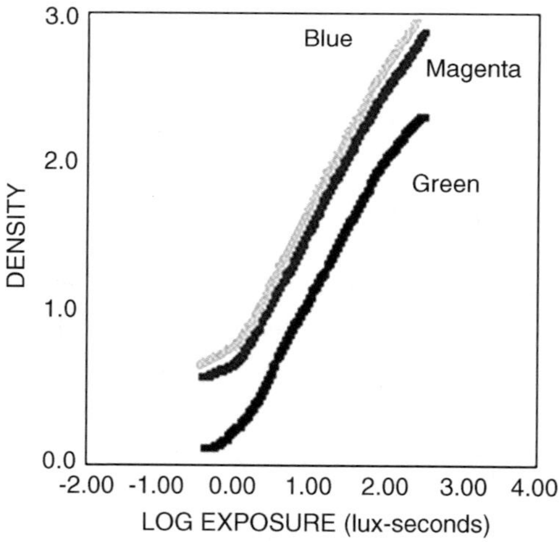

Figure 4 *Density vs log exposure for an intermediate film*

intermediate positive film and Figure 5, a typical black and white positive intermediate film, where the various curves in Figure 5 represent the different processing times of 4 min (curve 5), 5 min (curve 4), 6.5 min (curve 3), 9 min (curve 2) and 12 min (curve 1), respectively.

12.3 Print Film

The sensitometry of a typical colour print film is shown as Figure 6.

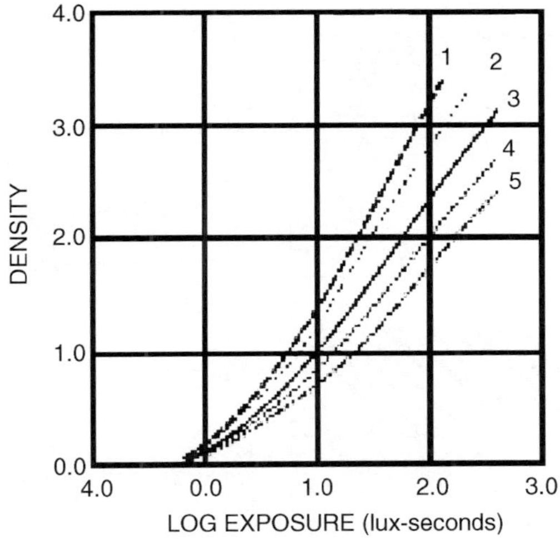

Figure 5 *A black and white intermediate film sensitometry at various processing times*

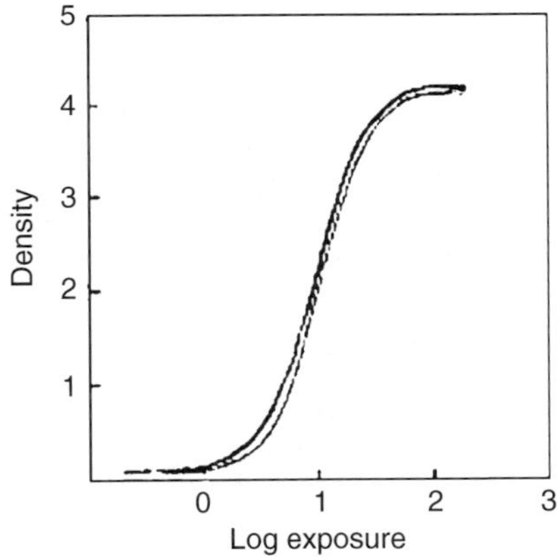

Figure 6 *Density vs log exposure for print film*

Unusually, a typical colour print film does not have the film structure of colour negative films, see Chapter 9. Figure 7 shows that the blue sensitive layer is the closest imaging layer to the film base, and the green sensitive layer is the uppermost light sensitive layer.[2]

These various films and curve shapes have been reproduced here to demonstrate the wide, almost endless, variety of possible sensitometric (density *vs.* log exposure) and colour reproduction responses for films. The same is valid for paper, it is merely the fact that there are less conditions under which paper products are exposed and therefore alternative paper formats are not required by the market.

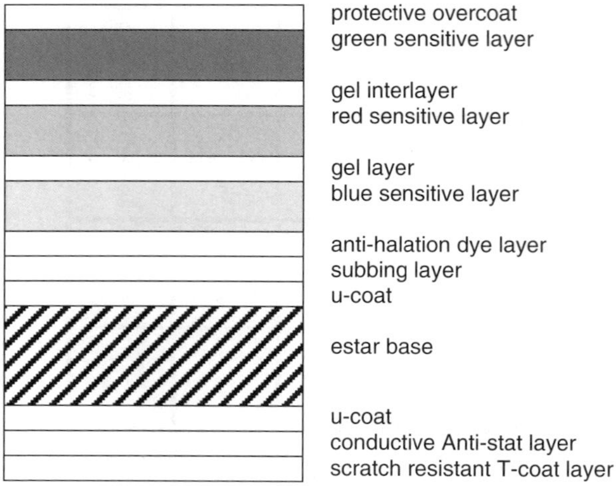

Figure 7 *Layer order for print film*

Component	Weight %
Water	95-99
Sodium carbonate	1-5
4- (N-ethyl-N-2-methanesulphonylaminoethyl) -2 - methylphenylenediamine sesquisulphate monohydrate	<1
Substituted phosphonate	<1
Sodium sulphite	<1

Figure 8 *Working strength developer solution for the ECN-2 processing kit*

Each film outlined above essentially uses chemistry outlined in previous chapters. They all use similar, although not identical, couplers from the coupler families described in Chapters 6 and 7.

The processing for these films uses a variety of processing solutions. The Eastman Kodak family (which at the time of writing had >95% of the Hollywood motion picture industry market share, and received more technical 'Oscars' than any other film manufacturer) uses the ECN processing solutions. The materials safety data sheet for ECN-2 kit colour developer replenisher lists the data in Figure 8 as those chemicals needed for the working solution.

The developer, 4- (*N*-ethyl-*N*-2-methanesulfonylaminoethyl) -2 -methylphenylenediamine sesquisulfate monohydrate, is the standard developer that was referred to earlier in Chapter 4, Figure 2.

References

1. J.C. Brewer, J.T. Keech and J.F. Sawyer, US 5,561,012, Eastman Kodak Co.
2. H-1-2383t from www.kodak.com.

CHAPTER 13

Instant Colour Photography

The concept and mechanism of instant, sometimes known as rapid access or one-step photography, was first published in 1936.[1] The Polaroid Corporation developed these early concepts, which resulted in a black and white product in 1948, and a colour product in 1963. There are two possible mechanisms. Mobile image dyes that are immobilised on interaction with exposed silver halide, Figure 1 or the use of non-diffusing compounds which release the image dyes on interaction with exposed silver halide, Figure 2.

Mordanted dyes is a term well known to the dye industry and relates to the process of forming adative or charged bonds between a small mobile molecule and a polymer.

Under normal circumstances, the process outlined in Figure 2 would produce a negative of the original scene. A direct positive can be obtained by the use of silver halide emulsions, which are precipitated in such a way as to produce an internal latent image when they are exposed to light. This type of emulsion is referred to as a reversal emulsion. Reversal emulsions cannot be developed using conventional emulsions but will react with nucleating or fogging agents, which are mixed into the developer solution, Figure 3.

The initial colour image transfer product, which was developed by Rogers[2] who worked for the Polariod Corporation, involved the imagewise immobilisation of dye developer molecules in a peel apart system. This system required that the material was viewed from the same side of the base as the exposure. The subsequent Eastman Kodak Co. product was designed so that the material was exposed from one side and viewed from the other side with no requirement for a peel-apart mechanism, Figure 4.

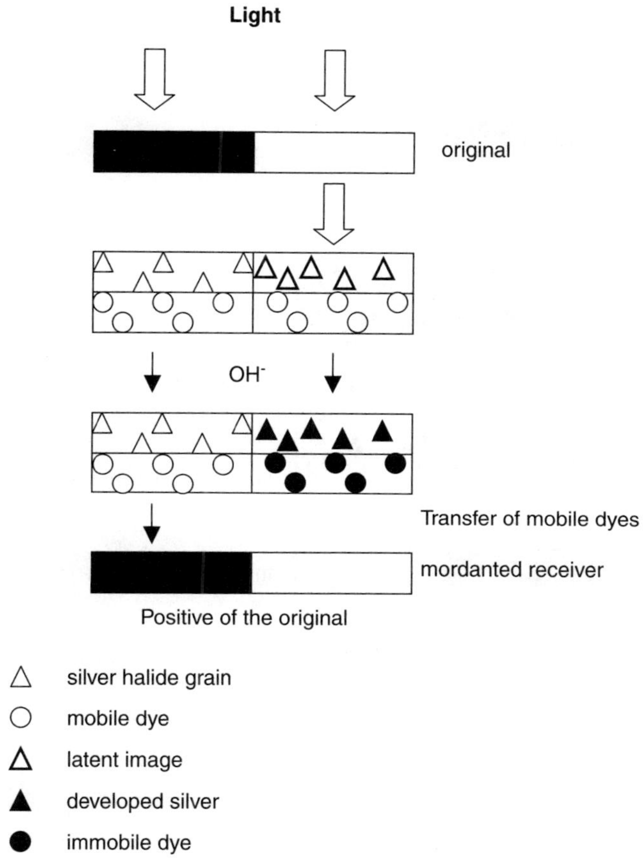

Figure 1 *The use of immobilised dyes*

13.1 SX-70

Figure 5 shows the SX-70 film structure in more detail.

The sensitivity of the silver halide layers is maintained by coating the image dye formers below the respective silver halide emulsion layers. A small mobile molecule, which is known as an electron transfer agent (ETA) is used to facilitate the reaction between the immobilised compounds and the externally fogged silver halide crystals. Phenidone and metol are two well-known ETAs, Figures 6 and 7, respectively.

The initial system designed by Rogers continued to be the basis of the Polaroid system, which involved the imagewise immobilisation of dye developer molecules. In the latent image areas, the hydroquinone, is oxidised to the quinone which is rendered immobile, Figure 8.

In the non-image areas the dye developer remains unaffected. By virtue of its solubility in the alkali activator solution, the dye developer

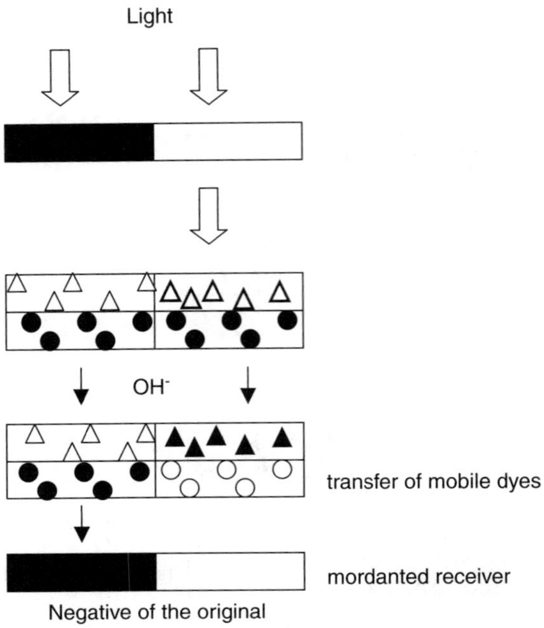

Figure 2 *The use of redox dye releasing couplers*

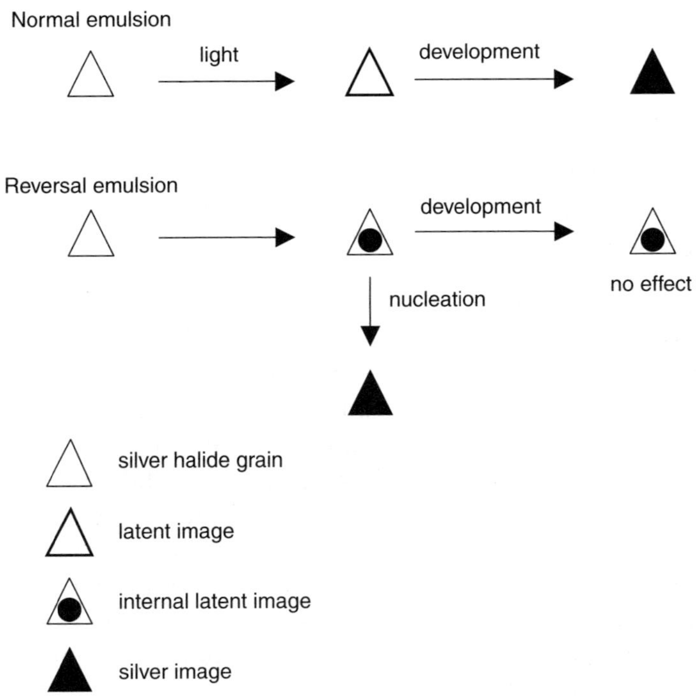

Figure 3 *A comparison of a normal and reversal emulsion*

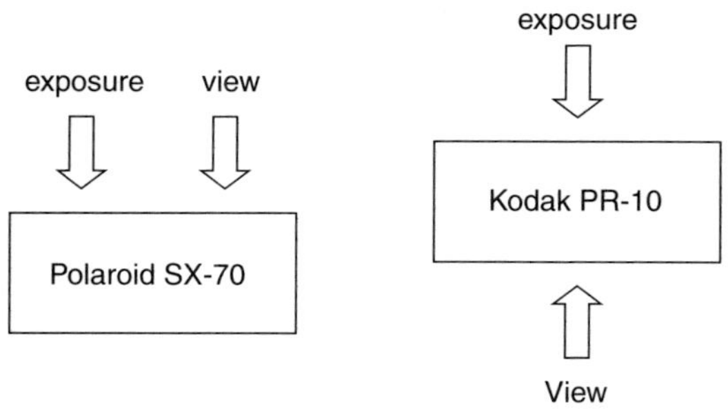

Figure 4 *A comparison of the two leading products of their day*

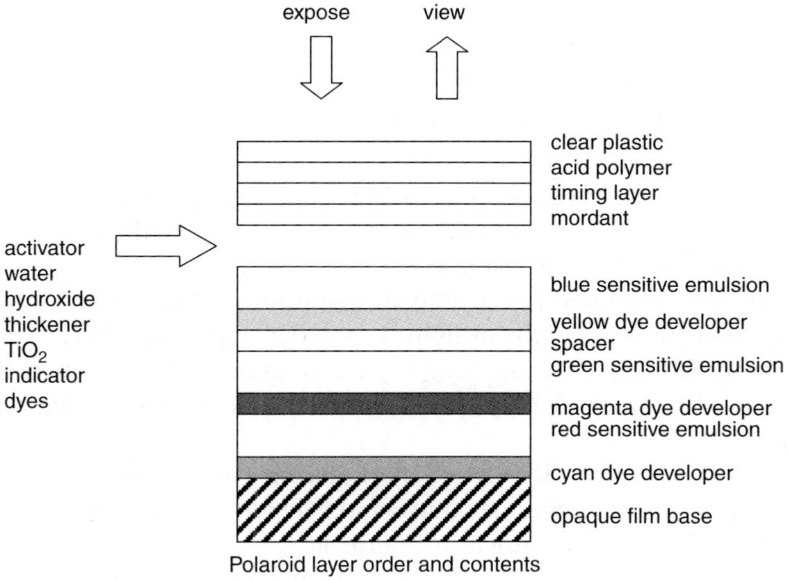

Figure 5 *The layer structure for SX-70*

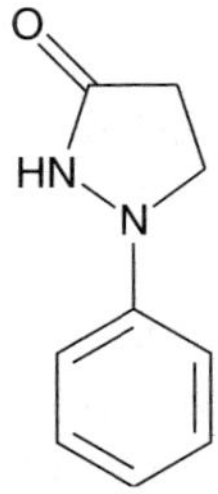

Figure 6 *The structure of phenidone*

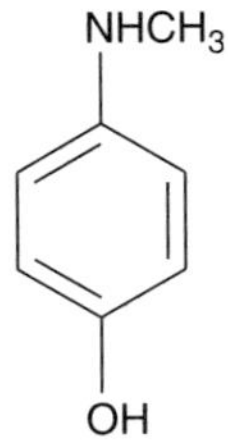

Figure 7 *The structure of metol*

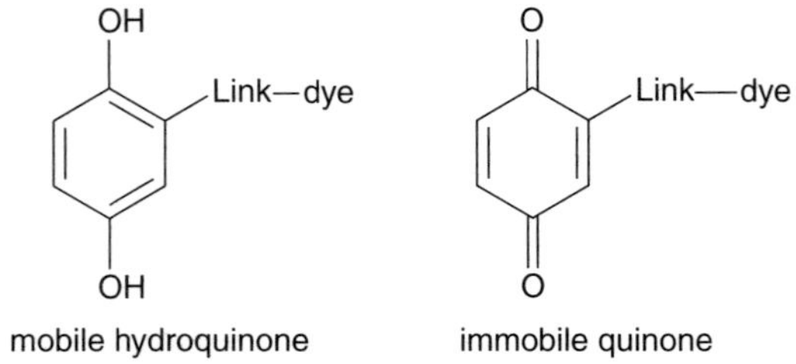

Figure 8 *Dyes are immobilised by converting hydroquinones to quinones*

will migrate to the receiving sheet where it is rendered immobile by the mordant, producing a positive image. Figure 9 details some of the reported Polariod dye developers.

Filter dyes are added to the activator solution so that the silver halide is protected from the ambient light during development. These dyes are often phthalein dye. In this case Figure 10 shows two dyes with λ_{max} values at 470 nm and 620 nm, respectively, in potassium hydroxide solution.

In his 1958 patent,[2] Rogers of the Polaroid Corporation reports the potential use of a number of compounds, which were suitable developers, some of which are shown in Figure 11.

The mordanted dyes using the commercial developer solutions require dye absorption curves as shown in Figure 12.

13.2 PR-10

A schematic diagram of the Eastman Kodak Co. system appears as Figure 13.

The Kodak system, first introduced in 1976, was based on the use of redox dye releasers (RDRs), which are intrinsically immobile in alkali. The exposed silver halide is developed by an ETA in a basic medium, producing oxidised ETA (ETA ox), which reacts with the RDR generating oxidised RDRs and regenerating the ETA. The oxidised RDR

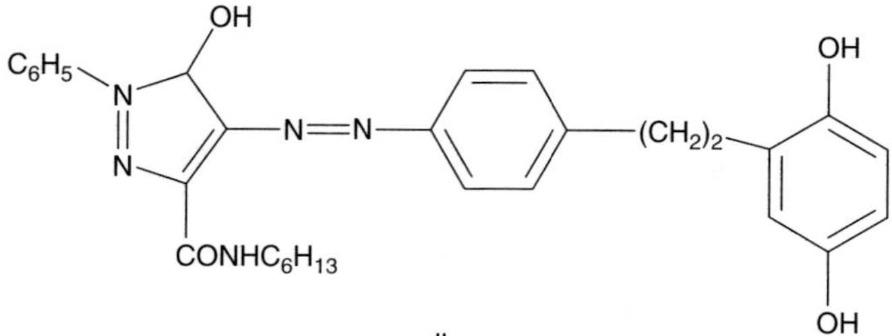

yellow

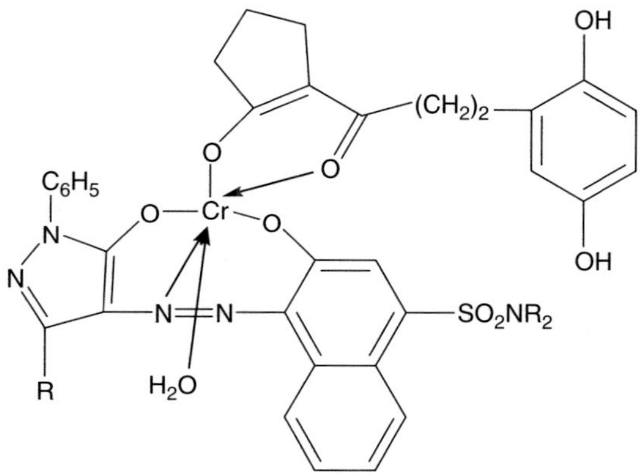

magenta

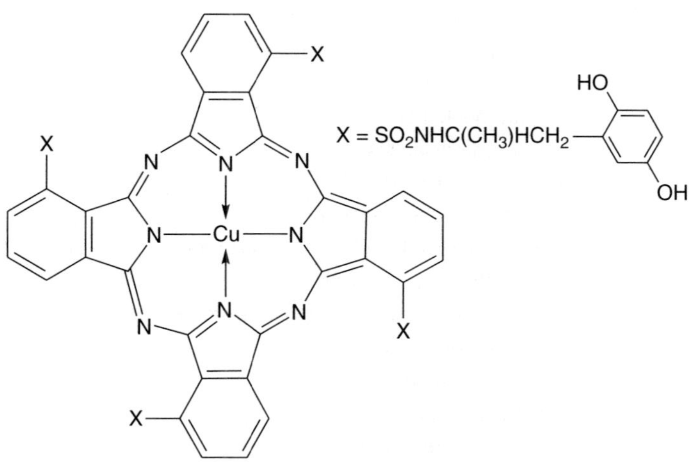

cyan

Figure 9 *Dye developers used by the Polaroid Corporation*

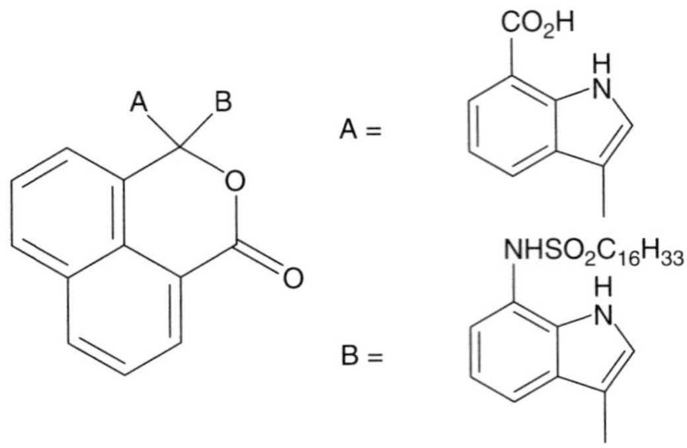

470 nm

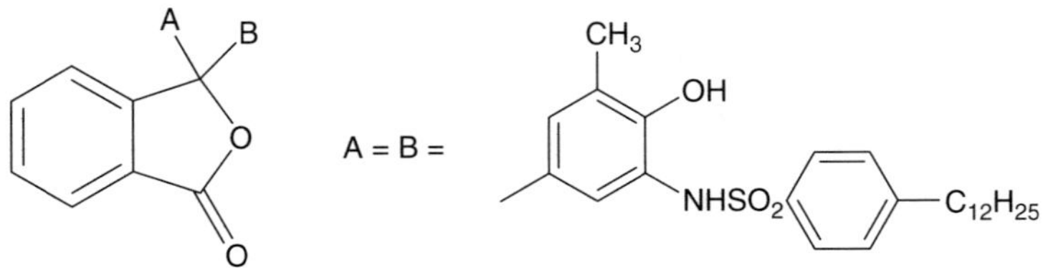

470 nm

Figure 10 *Filter dyes used in the SX-70*

- 2,4-diaminophenol
- toluhydroquinone
- 2,5-bis-ethyleneimino-hydroquinone
- p-benzylaminophenol
- p-anilinophenol
- xylohydroquinones
- p-toluthio-hydroquinone
- 2-amino 4-phenyl-phenol
- 5,8-dihydro-1,4-naphthohydroquinone
- o-toluthio-hydroquinone
- 5,6,7,8-tetrahydro-1,4-naphthohydroquinone
- 2-methoxy-4-aminophenol
- phenylhydroquinone
- p-aminophenylhydroquinone
- 4-amino-3-ethylphenol
- 4-amino-3,5-dimethylphenol
- 4-amino-2,3-dimethylphenol
- 4-amino-2,5-dimethylphenol
- 6-aminothymol
- thymolhydroquinone

Figure 11 *Examples of developers published by the Polaroid Corporation*

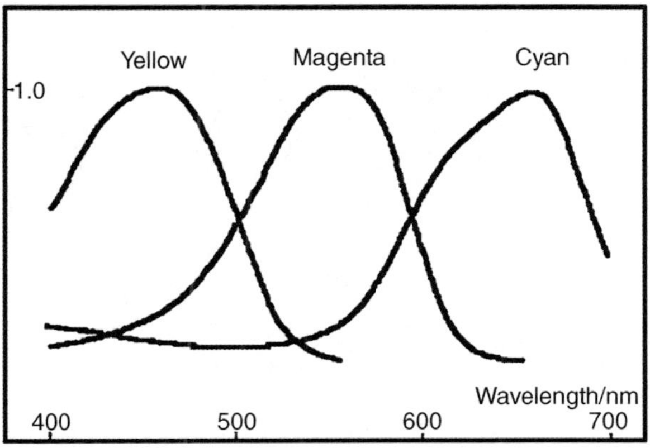

Figure 12 *Aim curve shapes for commercial developer solutions*

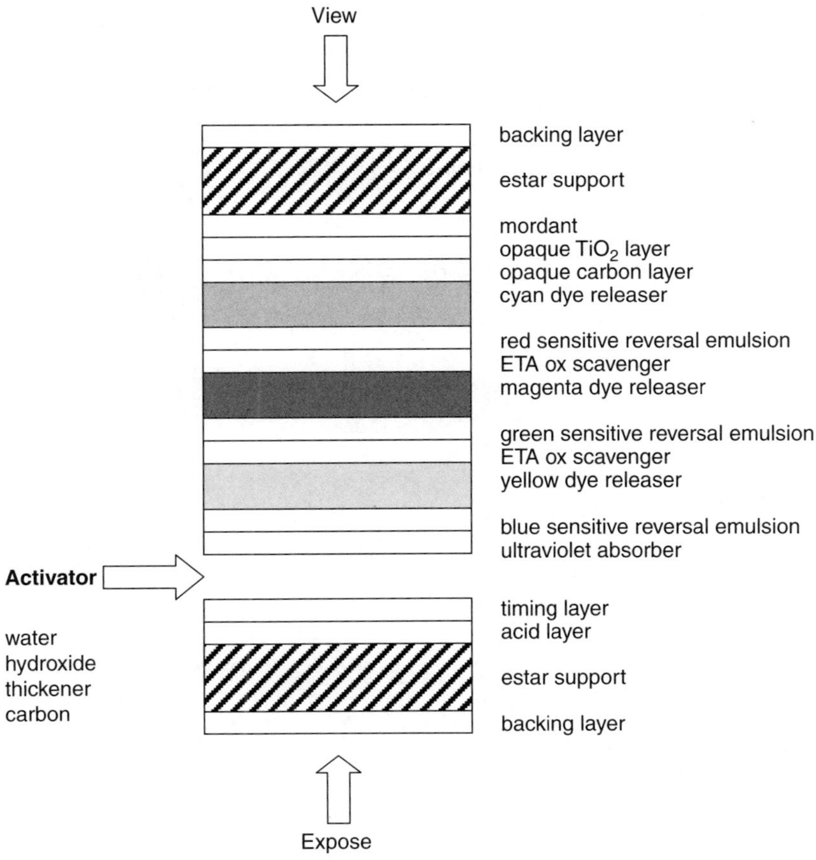

Figure 13 *The layer structure of PR-10*

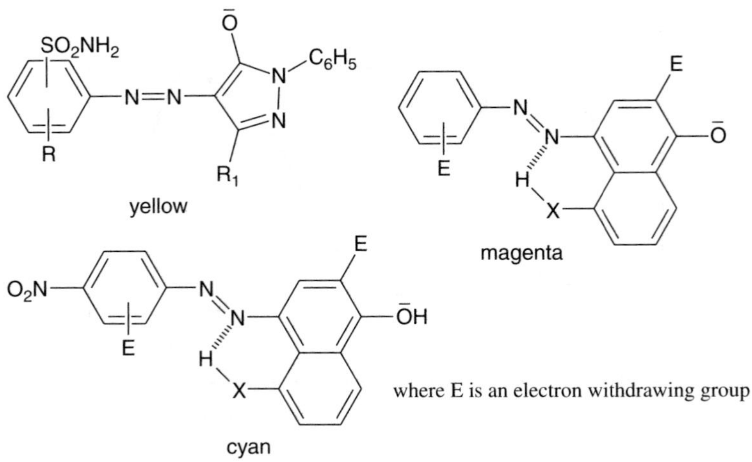

Figure 14 *Typical RDRs used in PR-10*

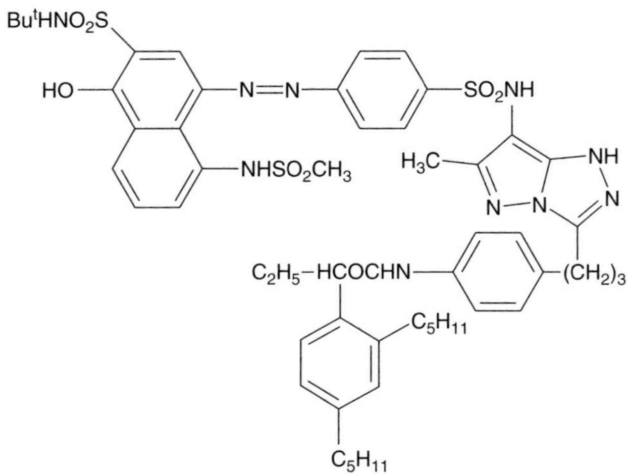

Figure 15 *A cyan RDR dye*

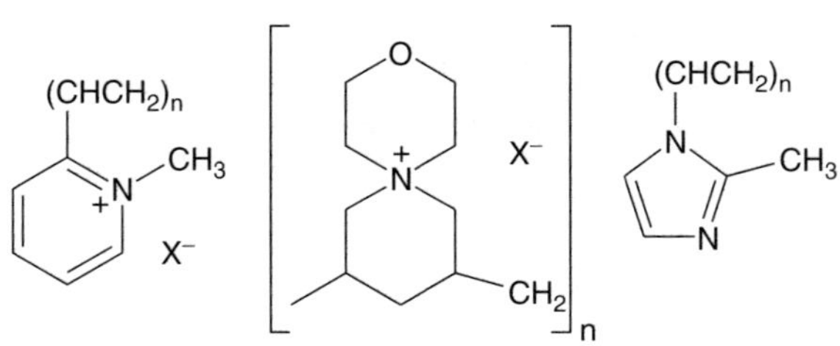

Figure 16 *Mordants of the type used in PR-10*

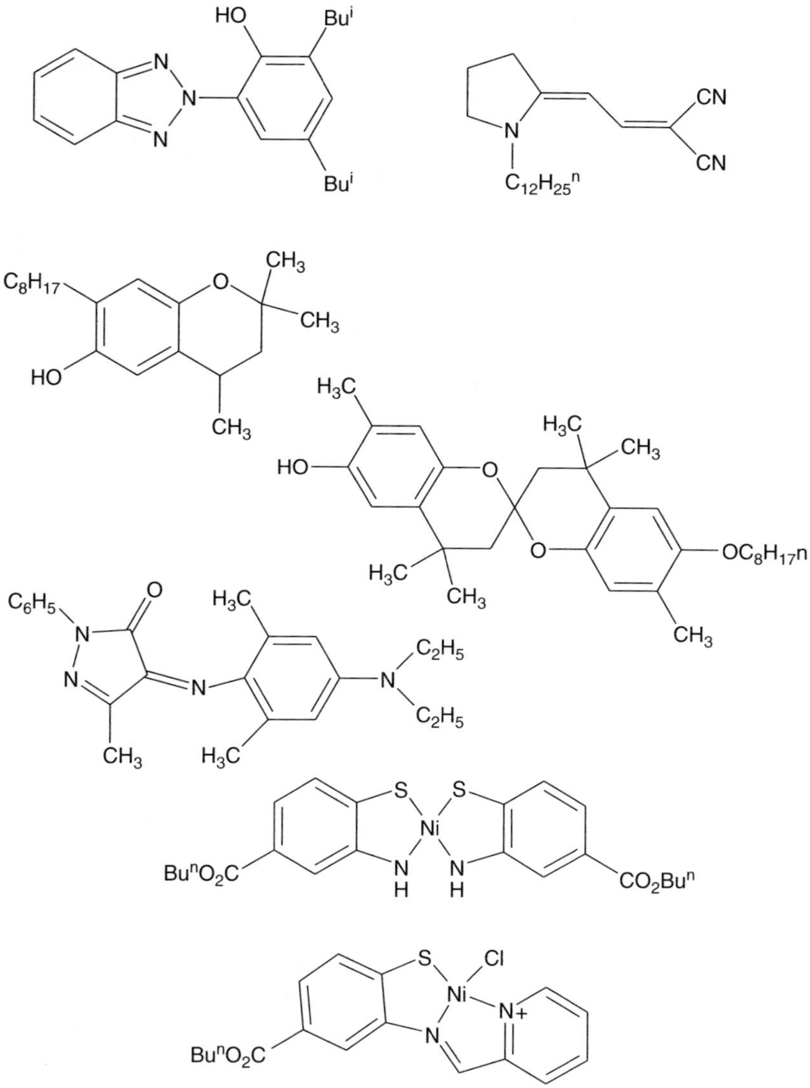

Figure 17 *Examples of known stabilisers used in CIT products*

readily hydrolyses in the basic medium, releasing a mobile dye, which migrates to the mordant layer. The production of a positive image requires the use of reversal emulsions as described above in Figure 3.

Bailey and Rogers[3] reported that the structure of the yellow RDRs was an arylazo derivative of a 5-pyrazolone, the magenta dye was an anionic derivative of a 4-arylazo-1-naphthol and the cyan dye was an azo derivative, Figure 14.

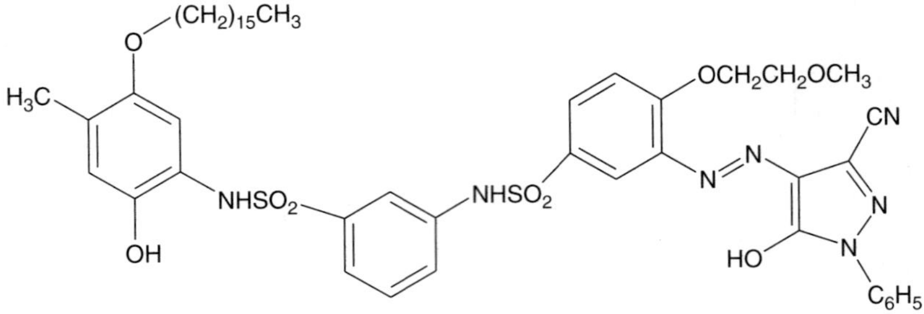

yellow

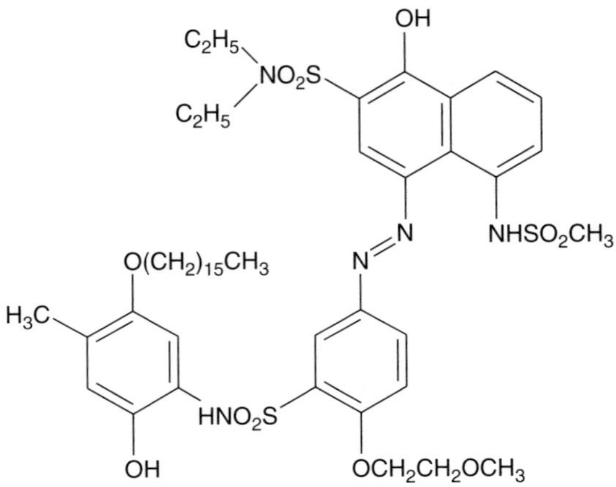

magenta

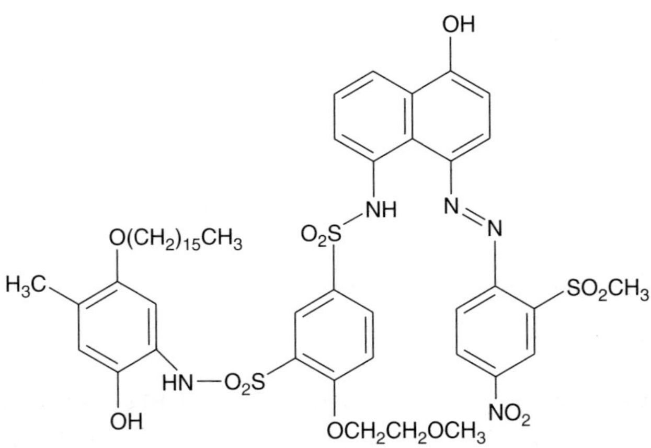

cyan

Figure 18 *Some dye releasing compounds used in the F 1-800G Fuji system*

Bailey and Clark[4] detailed the full structure of an example of a cyan RDR dye, in their report of reprographic techniques, Figure 15.

Mordants used for these types of RDRs need to be basic as the dyes tend to be acidic. They are usually polymeric so that they are immobile in their coated layer. Figure 16 outlines three useful mordants based on the pyridinium, morpholinium and imidazole nuclei, respectively.

UV protection of the image dyes has required the design and synthesis of stabilisers that are unique to the colour image transfer products. Figure 17 shows some of the reported compounds used as stabilisers.[4]

The systems described above which were produced by the Polaroid Corporation and Eastman Kodak Co. were not the only commercially available examples of image transfer materials. The Fuji Photo Film Company marketed a product called F10 Fotorama, launched in 1981. It was compatible with the Kodak PR-10 system. The Fuji system was complemented by a further product called the F 1-800G peel-apart system. These systems also used the concept of negative working emulsions and dye release chemistry. Figure 18 shows some of the dye releasing compounds from the F 1-800G Fuji system.[5]

References

1. K. Schinzel and L. Schinzel, *Photogr. Ind.*, 1936, **34**, 942.
2. H.G. Rogers, US 2,983,606, Polaroid Corporation.
3. J. Bailey and D.N. Rogers, *Chemistry in Britain*, London, March, 1985.
4. J. Bailey and B.A.J. Clark, In: *Comprehensive Heterocyclic Chemistry*, A.R. Katritzky and C.W. Rees (eds), vol 1, Chapter 14, 380, Pergamon Press Ltd, Oxford, 1984, ISBN (vol 1): 0-08-03701-9.
5. K. Keller (ed), *Science and Technology of Photography*, VCH, Weinheim, 1993, ISBN 1-56081-800-X, 137–142.

Section 2: The Chemistry of Digital Products

CHAPTER 14

Inkjet Paper

As we have seen, the manufacture of conventional silver halide films and papers was undertaken by only a few companies. These few manufacturers between them undertook all of the research and development, as well as manufacturing and product innovation. The advent of digital imaging systems has allowed manufacturers who were previously in other market segments to enter the business of marketing inkjet printing heads, ink and papers. While the increased number of manufacturers has resulted in a much wider choice for the consumer, there are some basic issues that are common, namely that the inkjet system provide

- lightfastness of an image,
- resistance to bleeding under high-temperature and high-humidity conditions, and
- resistance to indoor fading and discolouration of an image.

The process relies on the production of extremely small droplets that are formed from a stream of liquid ink, which owes its origin to Lord Rayleigh in the 1870s. The smallest droplets on the market are in the $1–10 \times 10^{-12}$ l, or 1–10 pl. Droplet sizes this small leads to very sharp images, which can be from a few millimetres to large display posters. This versatility requires a very sophisticated delivery system, *i.e.* the print head and potentially large volumes of inks.

Ink droplets are printed onto inkjet media, which ranges from clear plastic to highly sophisticated bespoke papers. The nature of the printers that are used varies with the application, for example

- continuous printing – typically industrial
- drop-on-demand – could use piezo or thermal printing heads.

14.1 Printing Inks

Standard or traditional inkjet inks have contained de-ionised water, a water-soluble organic solvent and a soluble dye or an insoluble pigment. The use of an insoluble pigment requires the use of a dispersant and possibly a surfactant. These compounds may increase the viscosity of the resultant solution, which may impact on the ability of the nozzle to provide the required droplet size. One solution to the problem is to consider the use of self-dispersing pigments, which are produced by modifying the surface of the pigment.

The standard printing ink dye set is yellow, magenta, cyan and black, see for example ref 1. There are some applications which use grey scale inks, and others that use some key colours, but these are in the minority. Examples of dyes that have been used include

Yellow:	Acid Yellow 17, Acid Yellow 23, Direct Yellow 86, Direct Yellow 123, Direct Yellow 132, Pigment Yellow 74, Yellow 1189, Yellow 104.
Magenta:	Acid Red 37, Acid Red 52, Acid Red 289, Pigment Red 122, Reactive Red 23 and Reactive Red 180.
Cyan:	Direct Blue 199, Acid Blue 9.
Blue:	Acid Blue 83, Acid Blue 260, Reactive Blue 19, Reactive Blue 21, Reactive Blue 49, Reactive Blue 72.
Green:	Reactive Green 12.
Red:	Acid Red 51, Acid Red 52, Acid Red 289, Acid Red 337, Acid Red 415, Reactive Orange 13, Reactive Orange 16, Reactive Red 43, Reactive Red 123.
Black:	AK 172, AK 194, DK 31, DB 195, Food Black 2, Monarch ® 880 (manufactured by the Cabot Corp).

Inks may also be formed from a mixture of dyes, for example, a red ink may be a mixture of Reactive Red 180 and Reactive Yellow 84, and green ink may be a mixture of Reactive Blue 72 and Reactive Yellow 85. The above list of published inkjet inks is not exhaustive. It does serve to illustrate the many choices open to the industry and also the vast potential of combinations of the above dyes.

Earlier discussions concerned the relationship between speed, grain and sharpness in the design of a colour film. Perhaps an equivalent triangle for inkjet inks may be provided as Figure 1.

The Lexmark Corporation suggested a typical formulation for black inkjet ink in their 2005 patent,[2] Figure 2.

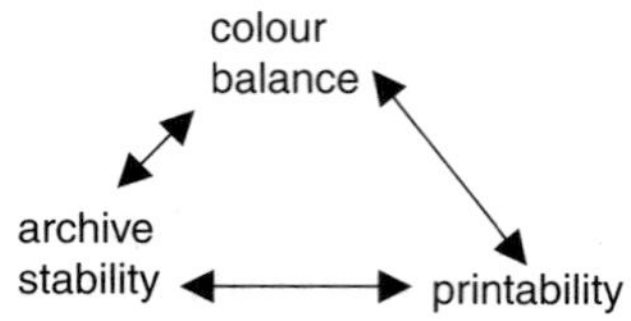

Figure 1 *Three major issues facing the inkjet manufacturers*

Component	% weight/weight
Monarch® 880 Pigment dispersed in a siloxy polymeric dispersant in a 5:1 pigment to dispersant ratio	2.0
Self-dispersing carbon black pigment	2.0
Polyethylene glycol (MW 400)	7.5
2-Pyrrolidone	7.5
l,2-Hexanediol	1.2
Hexylcarbitol	0.4
Binder (butyl acrylate/methyl methacrylate/acrylic acid)	2.0
Amino-containing compound	0.25-0.50
Water	balance

Figure 2 *A Lexmark Corporation formula for black ink*

In some cases a biocide is added. In their 2004 patent, Zimmer et al.[3] provide some of the suggested ink sets and dye sets. An ink set is a combination of dyes and/or pigments that produce a specific colour. A dye set is the combinations of ink sets that are used to generate all of the relevant colours. In the aforementioned dye set Lexmark report the use of yellow, magenta and cyan which are the three traditional dye sets used in printing that are most familiar to the amateur printing market.

Zimmer *et al.* suggest that the yellow ink set is a mixture of Acid Yellow 17, Acid Yellow 23, Direct Yellow 123, in addition to one of the three dyes listed in Figure 3. Not all of the yellow ink sets will contain the materials listed, but at least one of them will be present in each of their formulations.

The magenta dye set is listed as being Acid Red 52, Acid Red 289 or a combination in addition to one of the compounds listed as Figure 4.

The metals used in the above examples for the acid counter ion can be sodium, lithium, potassium or ammonium. Potential metals for the chelated ion in the left hand structure can be copper, nickel, iron or chromium.

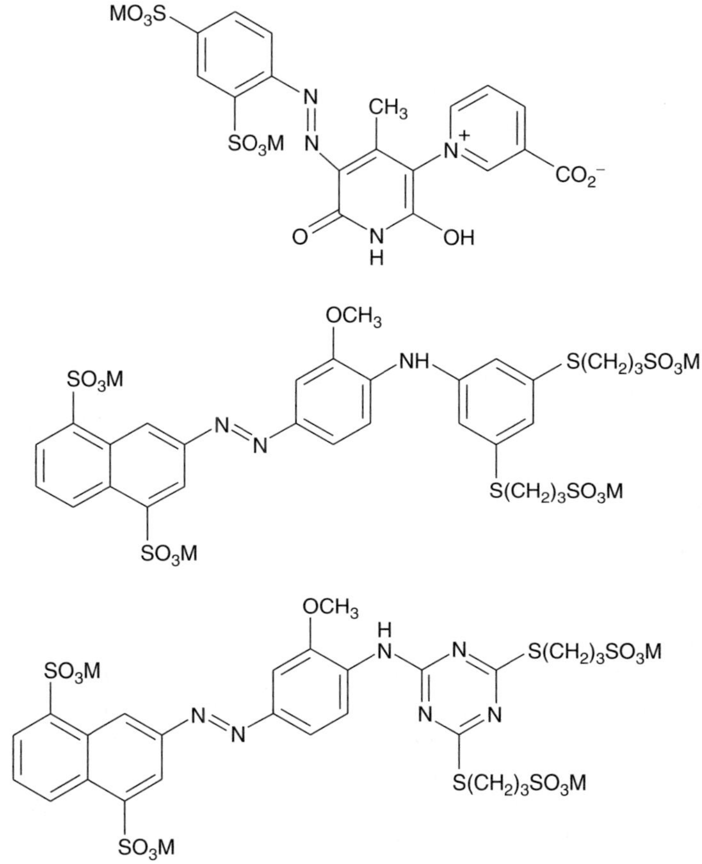

Figure 3 *Additional yellow inkjet inks*

The ratios of the magenta dyes (by total weight) are as follows:

Acid red 52	0–1.5%
Acid red 289	0–4%
One of the compounds in Figure 4	2–4%

The cyan ink is typically a mixture of Direct Blue 199 and Acid Blue 9. Zimmer *et al.* also mention the use of possible humectants as being

. . . *Humectants that may be employed in this invention are generally not limited and are known in the art. Illustrative examples include alkylene glycols, polyols, diols, bishydroxy terminated thioethers, and lactams. The alkylene glycols useful as humectants generally have a molecular weight of*

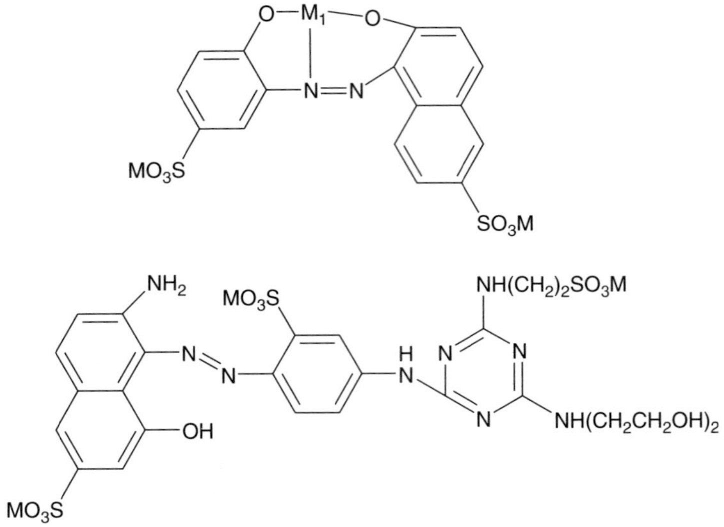

Figure 4 *Additional magenta inkjet inks*

from about 50 to about 4,000, preferably from about 50 to about 2,000, more preferably from about 50 to about 1,000, even more preferably from about 50 to about 500. Suitable polyalkylene glycols include polyethylene glycol, diethylene glycol, triethylene glycol, tetraethylene glycol, polyproylene glycol, dipropylene glycol, tripropylene glycol, and tetrapropylene glycol. In another embodiment, the humectant comprises tetraethylene glycol and tripropylene glycol . . .

They go on to mention the potential range of penetrants as:

. . . Penetrants that may be employed in this invention are generally not limited and include hydroxy substituted hydrocarbons like 1,2-alkyl diols such as 1,2-pentanediol, 1,2-hexanediol and mixtures thereof. A more detailed description of such penetrants maybe found in U.S. Pat. No. 5,364,461. Additional examples of penetrants include: alkyl alcohols having 1 to 4 carbon atoms, such as ethanol, methanol, butanol, propanol and isopropanol; glycol ethers, such as ethylene glycol monomethyl ether, ethylene glycol monoethyl ether, ethylene glycol monobutyl ether, ethylene glycol monomethyl ether acetate, diethylene glycol monomethyl ether, diethylene glycol monoethyl ether, diethylene glycol mono-n-propyl ether, ethylene glycol mono-iso-propyl ether, diethylene glycol mono-iso-propyl ether, ethylene glycol mono-n-butyl ether, ethylene glycol-mono-t-butyl ether, diethylene glycol mono-t-butyl ether, 1-methyl-lmethoxy butanol, propylene glycol monomethyl ether, propylene glycol monoethyl ether, propylene glycol mono-tbutyl ether, propylene glycol mono-n-propyl ether, propylene glycol mono-iso-propyl ether, dipropylene glycol monomethyl

ether, dipropylene glycol monoethyl ether, dipropylene glycol mono-n-propyl ether, and dipropylene glycol mono-iso-propyl ether; formamide; acetamide; dimethylsulfoxide; sorbitol; sorbitan; acetin; diacetin; triacetin; and sulfolane. A preferred penetrant is 1,2-hexanediol . . .

The patent text also provides suggested amounts for their potential use

. . . The amount of penetrant in the ink composition may range from about 0.01 to about 20% by weight. In an embodiment, the amount of penetrant ranges from 0.1 to 10% by weight. In another embodiment, the amount of penetrant ranges from about 0.5 to about 5% by weight . . .

Other materials that might be used in the formulation include chelating agents and surfactants.

There are applications where a near infrared dye is printed, for example, for providing markings for security/authentification, sorting, *etc*. While these types of dyes cannot be seen by the human eye, they can be detected by instruments. These dyes have strong absorptions in the 700–900 nm range. Unfortunately, some cyan dyes do have overlapping dye absorptions precluding the use of some of these dyes. Under these circumstances, the dye set can be blue, green, magenta and yellow.[1]

The inks that are used can produce invisible printing/marks on a variety of surfaces including cellophane, fabric, film sheeting, glass, high gloss paper, metal foils, paper, plastic, plates, rubber and wood. The ink formulation contains an organic solvent in which is dissolved a soluble polymer, which is composed of an infrared fluorophore polymerised with a soluble polyester. The solvents include[2]

- carbon ketones (C_3-C_6)
- carbon organic ester (C_3-C_6)
- an alcohol
- a combination of one or more of the above.

In their detailed description of their invention, Escano *et al.*[4] comment of the inks, which can be used for inkjet printing using the piezoelectric impulse drop-on-demand method and single nozzle inkjet printers (p. 3).

a. between about 1 and about 10 weight percent of at least one non-sulfo containing, organic solvent soluble polyester having from about 0.1 ppm by weight to about 10% by weight of a thermally stable near infrared fluorophore copolymerized therein
b. between about 1 and about 5 weight percent of an organic solvent binder resin selected from a cellulose ester and/or condensed phenolic resin
c. between about 0.01 and about 0.5 weight percent of a corrosion inhibitor

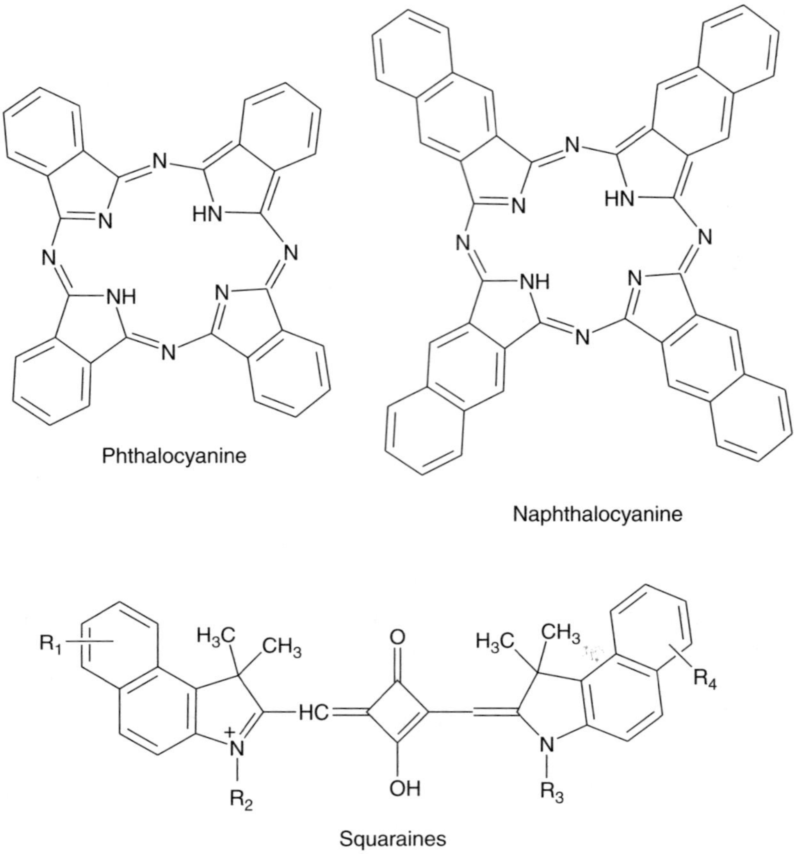

Figure 5 *Examples of near infrared fluorophores*

> *d. between about 0.50 and about 1.0 weight percent of an organic solvent soluble electrolyte*
> *e. and the remainder of said ink consisting of at least one C3 C6 aliphatic ketone, at least one C3-C6 aliphatic ester, or a combination thereof, with all of the weight percentages being based on the total weights of components a–e.*

Figure 5 highlights the parent compounds from three families of compounds of near infrared fluorophores from the phthalocyanines, naphthalocyanines and squaraines. In practice derivatives of these compounds will be used the exact formula for which may well be difficult to obtain.

14.2 Inkjet Media

Inkjet media or paper is commercially available in many formats, which differ in both complexity of the technical design and physical size formats. The ability to manufacture and process large format photographic products was a constraint that is no longer applicable for there is now no need for wet chemistry processing.

The technical complexity of the papers varies widely, as the applications vary. For example, there may be a need to print an image just as a proof and not be overly concerned with colour rendition or long-term archival stability. At the other extreme, there may be a need to produce the highest quality image with the expectation of archival stability of years and not days.

The least complex of the inkjet papers are those with compounds added to the wood pulp prior to paper manufacture, see for example ref 5. This type of paper may well represent the largest volume of paper sold for use in the amateur market and for colour photocopiers. The additives tend to be compounds such as succinic anhydride, silicic acid and cationised starch. A more detailed discussion of these types of papers should more rightly appear in a text concerning paper.

The more complex inkjet papers that deliver image quality close to if not equal to traditional photographic quality require several layers. The basic principles are outlined in Figure 6.

In the left hand diagram, the arrow has been used to signify the application of an inkjet droplet. The ink droplet contains ink plus solvent. The right hand diagram in Figure 6 shows that the ink is trapped in the dye trapping layer and the solvent passes through into the ink carrier liquid receptive layer. The solvent droplet is depicted here as light grey for demonstration purposes only. At this stage, the various layers are still porous. The final stage in the process requires either heat or pressure to render the layers impermeable.

These basic principles have then been used in more complex formulations to cater for some of the constraints of paper base. For example, there is still a requirement for a polymer layer between the ink carrier

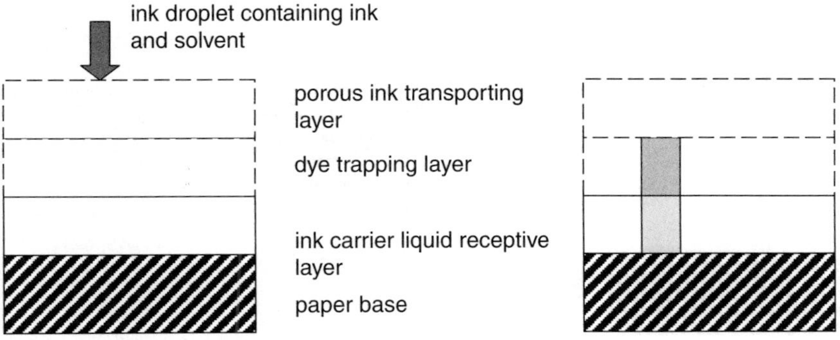

Figure 6 *The basic principles of inkjet paper*

liquid receptive layer and the paper base. Without this layer there would be the potential for the ink carrier to diffuse into the paper base.

In their 2006 patent, DeMejo *et al.*[6] provide some suggestions of the chemistry of these layers.

14.2.1 Ink Carrier Liquid Receptive Layer

The components were mixed and diluted to an 18% aqueous solution, which was then coated on the pre-prepared base to give a dry laydown of 8.6 gm^{-2} (Figure 7).

14.2.2 Dye Trapping Layer

Divinylbenzene-co-*N*-vinylbenzyl-*N*,*N*,*N*-trimethylammonium chloride is a mordant that is used to trap the dye (Figure 8). The mechanism is similar to the redox dye releasing materials discussed in Chapter 13. In their patent, DeMejo *et al.*[6] comment further on the types of suitable mordants thus:

> *. . . Such a dye mordant can be any material that is effectively substantive to the inkjet dyes. The dye mordant removes dyes from the ink received from the porous ink-transporting layer and fixes the dye within the dye-trapping layer. Examples of such mordants include cationic lattices such as disclosed in U.S. Pat. No. 6,297,296 and references cited therein, cationic polymers such*

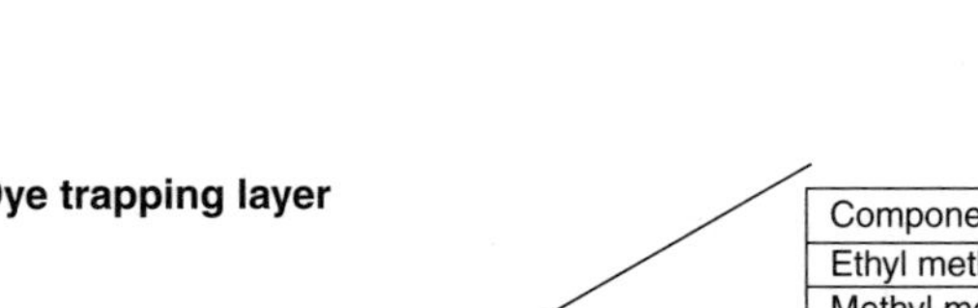

Component	Ratio
Ethyl methacrylate	95
Methacrylic acid	5

Figure 7 *Components of the inkjet carrier liquid receptive layer*

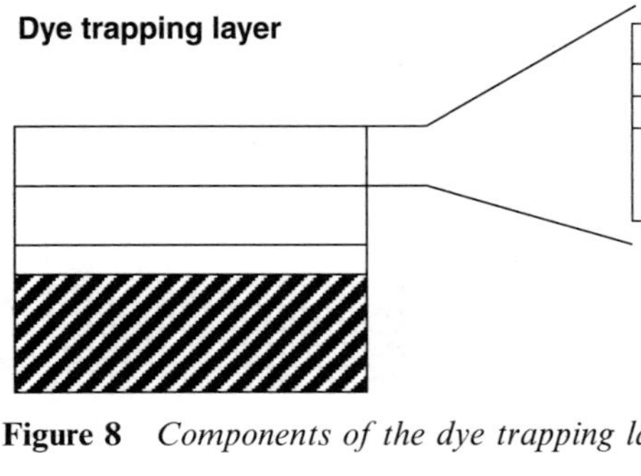

Component	Dry weight Ratio
Ethyl methacrylate	80
Methyl methacrylate	10
Divinylbenzene-co-N-vinylbenzyl-N,N,N-trimethylammonium chloride	10

Figure 8 *Components of the dye trapping layer*

as disclosed in U.S. Pat. No. 5,342,688, and multivalent ions as disclosed in U.S. Pat. No. 5,916,673, the disclosures of which are hereby incorporated by reference. Examples of these mordants include polymeric quaternary ammonium compounds, or basic polymers, such as poly(dimethylaminoethyl)-methacrylate, polyalkylenepolyamines, and products of the condensation thereof with dicyanodiamide, amine-epichlorohydrin polycondensates. Further, lecithins and phospholipid compounds can also be used. Specific examples of such mordants include the following: vinylbenzyl trimethyl ammonium chloride/ethylene glycol dimethacrylate; poly(diallyl dimethyl ammonium chloride); poly(2-N,N,N-trimethylammonium)ethyl methacrylate methosulfate; poly(3-N,N,N-timethyl-ammonium)propyl methacrylate chloride; a copolymer of vinylpyrrolidinone and vinyl(N-methylimidazolium chloride; and hydroxyethylcellulose derivatized with 3-N,N,N-trimethylammonium-)propyl chloride. In a preferred embodiment, the cationic mordant is a quaternary ammonium compound . . .

14.2.3 Ink Transporting Layer

For ink transporting layer, see Figure 9.

DeMejo *et al.*[6] commented:

. . . Upon fusing of the polymeric particles, the air-particle interfaces present in the original porous structure of the layer are eliminated and a non-scattering, substantially continuous, protective overcoat forms over the image. In a preferred embodiment of the invention, the fusible, polymeric particles in the ink-transporting layer comprise a cellulose ester polymer, such as cellulose acetate butyrate, a condensation polymer, such as a polyester or a polyurethane, or an addition polymer, for example, a styrenic polymer, a vinyl polymer, an ethylene-vinyl chloride copolymer, a polyacrylate, poly(vinyl acetate), poly(vinylidene chloride), and/or a vinyl acetate-vinyl chloride copolymer. In a preferred embodiment of the invention, the fusible, polymeric particles are comprised of a polyacrylate polymer or copolymer (for example, acrylic beads) comprising one or more monomeric

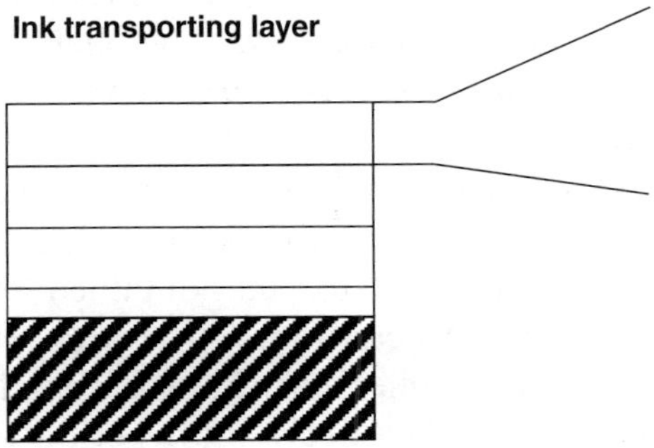

Component
Cellulose acetate butyrate
A condensation polymer, eg a polyacrylate
Optional binder

Figure 9 *Components of the ink transporting layer*

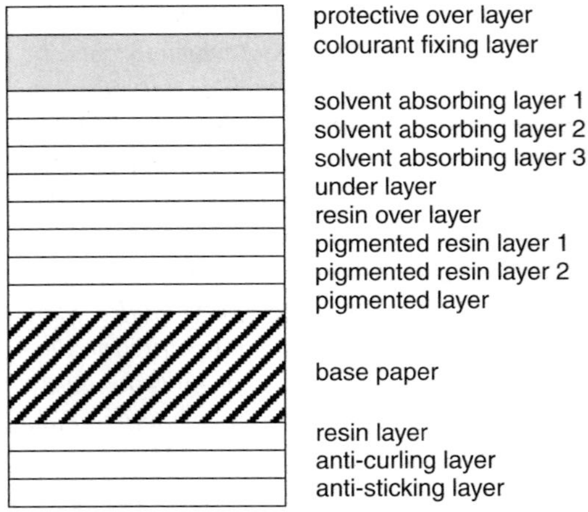

Figure 10 *An example of a Fuji Photo Film Co inkjet paper layer structure*

units derived from an alkyl acrylate or alkyl methacrylate monomer, wherein the alkyl group preferably has 1 to 6 carbon atoms . . .

The above example is but one formulation from one of the inkjet paper manufacturers. The Fuji Film Co. provide a layer structure of one of their inkjet papers on their website which details many layers,[7] Figure 10.

The principles described above are still valid with this Fuji Film Co. formulation as there is a layer which mordants the dye, which is known as the colourant-fixing layer. In this case there are three solvent absorbing layers. The use of three solvent absorbing layers will allow for the trapping of a great deal of solvents that could result from the application of large amounts of ink per square centimetre. The pigmented layers allow for the use of materials such as titanium dioxide, which is well known in the manufacture of photographic papers and is used to help reflect the light from the paper base. Anti-curl layers often contain the same substrate as is coated on the other side of the base. They help to keep dimensional stability when using large sheets, which are more prone to exhibit issues relating to curl.

In relative terms the use of products specifically designed to print images from digital sources is in its infancy. Nevertheless under some circumstances the image quality is comparable to that obtained by conventional wet chemistry. It is difficult to determine if the archival stability of these images will stand the test of time, but what is certain is that the technology has been developed in a relatively short period of time to produce outstanding products from all of the manufacturers.

References

1. C. Jackson, WO 2006/028910, E. I Dupont de Nemours and Company.
2. C. E. Akers, WO 2005/010105, Lexmark International Inc.
3. A. Zimmer, J.M. Medley, V. Kantrovich, W. Lake and S. McCain, US 2004/0123772, Lexmark International Inc.
4. N.Z. Escano, J.J. Krutak and M.A. Weaver, US 5,990,197, Eastman Kodak Co.
5. T. Ogino, K. Hosoi, C. Koga and T. Matsuda, US 2004/0121093, Fuji Xerox Co. Ltd.
6. L.P. DeMejo, X. Wang, G.E. Missell and A. Wexler, US 2006/0003112, Eastman Kodak Co.
7. http://www.fujifilm.com/jsp/fuji/epartners/.

Bibliography

1. R.J. Cox (ed), *Photographic Gelatin II*, Academic Press Inc., London, 1976, ISBN 0-12-194452-2.
2. K. Keller (ed), *Science and Technology of Photography*, VCH, Weinheim 1993, ISBN 1-56081-800-X.
3. L.F.A. Mason, *Photographic Processing Chemistry*, 1975, Focal Press Ltd. London, ISBN 0-240-50824-6.
4. T.H. James, *The Theory of the Photographic Process*, 4th edn, Macmillan Publishing Co Inc, New York, 1977, ISBN 0-02-360190-6.
5. *Journey: 75 Years of Kodak Research*, 1989, printed by Eastman Kodak Co., ISBN 0-87985-653-X. http://wwwsy.kodak.com/global/en/professional/support/databanks/filmDatabankReference.jhtml.
6. R.W.G. Hunt, *The Reproduction of Colour*, 5th edn, 1995, Fountain Press, Kingston-upon-Thames, ISBN 0 86343 381 2.
7. P. Kowaliski, *Applied Photographic Theory*, 1972, Wiley, Vincennes, ISBN 0 471 50600 1.
8. J.M. Sturge, V. Walworth and A. Shepp, *Imaging Processes and Materials*, 1989, 8th edn, Van Nostrand Reinhold, New York, ISBN 0-442-28042-6.
9. B.H. Carroll, G.C. Higgins and T.H. James, *Introduction to Photographic Theory – The Silver Halide Process*, 1980, Wiley, Chichester, ISBN 0-471-02562-3.
10. H. Asher, *Photographic Principles and Practices*, 1970, Fountain Press Ltd., London, ISBN 852 42170 2.
11. E.S. Bomback (ed and revised), *The Science of Photography*, 1967, Oxley Press Ltd., Nottingham, ISBN 852-42210-5
12. L. Stroebel, J. Compton, I. Current and R. Zakia, *Photographic Materials and Processes*, 1986, Focal Press, London, ISBN 0-240-51752-0.
13. C.E.K. Mees, *From Dry Plates to Ektachrome Film*, 1961, Ziff-Davis Publishing Company., New York.

14. T. Tani, *Photographic Sensitivity*, 1995, Oxford, University Press, Oxford ISBN 0-19-507240-5.
15. A. Mortimer, *Colour Reproduction in the Printing Industry*, 1991, PIRA International, Surrey, ISBN 0-902799-76-2.
16. J.A.C. Yule, *Principles of Colour Reproduction*, 1967, Wiley, New York, Library of Congress Card number 66-26764.

Subject Index